Enhancing the Mission Command Training of Army Functional and Multifunctional Brigade Headquarters for Large-Scale Combat Operations

JOSHUA KLIMAS, JENNIFER KAVANAGH, DEREK EATON

Prepared for the United States Army
Approved for public release; distribution unlimited

For more information on this publication, visit **www.rand.org/t/RR3260**.

About RAND

The RAND Corporation is a research organization that develops solutions to public policy challenges to help make communities throughout the world safer and more secure, healthier and more prosperous. RAND is nonprofit, nonpartisan, and committed to the public interest. To learn more about RAND, visit www.rand.org.

Research Integrity

Our mission to help improve policy and decisionmaking through research and analysis is enabled through our core values of quality and objectivity and our unwavering commitment to the highest level of integrity and ethical behavior. To help ensure our research and analysis are rigorous, objective, and nonpartisan, we subject our research publications to a robust and exacting quality-assurance process; avoid both the appearance and reality of financial and other conflicts of interest through staff training, project screening, and a policy of mandatory disclosure; and pursue transparency in our research engagements through our commitment to the open publication of our research findings and recommendations, disclosure of the source of funding of published research, and policies to ensure intellectual independence. For more information, visit www.rand.org/about/principles.

RAND's publications do not necessarily reflect the opinions of its research clients and sponsors.

Published by the RAND Corporation, Santa Monica, Calif.
© 2022 RAND Corporation
RAND® is a registered trademark.

Library of Congress Control Number: 2022905957
ISBN: 978-1-9774-0439-8

Limited Print and Electronic Distribution Rights

Preface

This report documents research and analysis conducted as part of a project entitled *Training Readiness for Headquarters of Separate Brigades*, sponsored by U.S. Army Forces Command. The purpose of this project was to analyze mission command training conducted by different types of brigade headquarters. This analysis was aimed to identify gaps in existing training; ways that these gaps could be overcome using current or revised training approaches; proposals for approaches to mission command training that can be applied forcewide and across types of brigade headquarters; and differences across brigade types that must be incorporated into the design and execution of training to support mission command.

This research was conducted within RAND Arroyo Center's Personnel, Training, and Health Program. RAND Arroyo Center, part of the RAND Corporation, is a federally funded research and development center (FFRDC) sponsored by the United States Army.

RAND operates under a "Federal-Wide Assurance" (FWA00003425) and complies with the *Code of Federal Regulations for the Protection of Human Subjects Under United States Law* (45 CFR 46), also known as "the Common Rule," as well as with the implementation guidance set forth in U.S. Department of Defense (DoD) Instruction 3216.02. As applicable, this compliance includes reviews and approvals by RAND's Institutional Review Board (the Human Subjects Protection Committee) and by the U.S. Army. The views of sources utilized in this study are solely their own and do not represent the official policy or position of DoD or the U.S. Government.

Contents

Figures and Tables

Summary

The research reported here was completed in September 2019, followed by security review by the sponsor and the Office of the Chief of Public Affairs, with final sign-off in December 2021.

Mission command involves the way in which "commanders, supported by their staffs, combine the art of command and the science of control to understand situations, make decisions, direct action, and accomplish missions."[1] This report examines the effectiveness of mission command training conducted by different types of functional and multifunctional (F/MF) brigade headquarters in preparation for large-scale combat operations (LSCO).[2] The U.S. Army's exercise of mission command as part of counterinsurgency (COIN) and stabilization operations during Operation Enduring Freedom (OEF) and Operation Iraqi Freedom (OIF) differs from the way that it would do so as part of LSCO. Army leaders have expressed concerns that the abilities of leaders and their staffs to exercise mission command as part of LSCO have atrophied. The U.S. Army Training and Doctrine Command's *U.S. Army in Multi-Domain Operations 2028* emphasizes that forces must be ready to deploy and *immediately* enter the fight in a fast-moving, contested, antiaccess and area denial environment—which places a premium on the quality and rigor of peacetime training.[3] RAND researchers' objective was to identify gaps in current training approaches for LSCO and recommend ways that these gaps could be overcome.

Overall, we find that the Army provides valuable mission command training opportunities for many F/MF brigade headquarters. However, various limitations mean that training falls short of a true gold standard for LSCO. The Warfighter Exercise (WFX)—a multi-echelon, simulation-driven command post exercise for units at the corps, division, and brigade headquarters levels—provides what is arguably the Army's premier mission training opportunity for F/MF brigade headquarters. WFX training surpasses what units can accomplish at home station in many ways. Joint and combined exercises also provide venues with significant training benefits, but these often complement, rather than substitute for, the training that a WFX provides.

However, the WFX has capability and capacity constraints that limit its effectiveness in meeting the training objectives of participating F/MF brigade headquarters. For example, the Army lacks the capacity to include all types of F/MF brigade headquarters as WFX training audiences. However, given resource constraints, we find that the types of F/MF brigade head-

---

[1]   Army Doctrine Publication 6-0, *Mission Command*, Change No. 2, Washington, D.C.: Headquarters, Department of the Army, March 12, 2014, p. ii.

[2]   This report focuses specifically on F/MF *brigade headquarters*. It *does not* assess training approaches for the various types of units that are subordinate to different F/MF brigade headquarters. Any conclusions that we might draw regarding gaps or challenges in training approaches for F/MF brigade headquarters are not intended to imply that comparable gaps and challenges necessarily apply at subordinate echelons.

[3]   U.S. Army Training and Doctrine Command, *The U.S. Army in Multi-Domain Operations 2028*, Fort Eustis, Va.: Headquarters, TRADOC Pamphlet 525-3-1, November 27, 2018.

quarters currently included as training audiences reasonably represent those that should derive the most benefit from WFX participation.

More significantly, the training objectives of F/MF brigade headquarters that are WFX training audiences are secondary to those of participating division and corps headquarters. The Mission Command Training Program is often unable to address some of the training objectives for F/MF brigade headquarters, as doing so would divert resources from the training objectives of the participating division and corps headquarters. In particular, *the WFX generally does not permit F/MF brigade headquarters to train to failure*, as this would risk the ability of divisions and corps to meet their training objectives. Moreover, some Mission Essential Tasks (METs) for some brigade types cannot be effectively trained at a WFX (or at home station); some form of field training or other specialized training is required. In short, although a WFX provides valuable training for the F/MF brigade headquarters that participate, it appears inadequate for preparing them to exercise mission command across the full range of wartime METs.

We found it difficult to assess the potential risk to force and mission resulting from these limitations. The F/MF brigade headquarters staff that we interviewed generally estimated that the risks were not significant and could be overcome in a relatively short period of time at the start of an operation. However, most personnel have focused on COIN operations since 2003, and they might be limited in their ability to fully assess risk for LSCO. Army doctrine states that units are supposed to receive external evaluations (EXEVALs) from higher headquarters as an objective way to evaluate training proficiency. However, none of the brigade headquarters staff that we interviewed reported receiving an EXEVAL of LSCO proficiency, although most indicated that they would like to receive one.

Because the unit staff we interviewed generally estimated that risks were not significant, we do not recommend wholesale changes to current F/MF brigade headquarters training approaches. Instead, we offer recommendations for prioritization and experimentation. We recommend that the Army *experiment* with an enhanced WFX that provides more time to focus on F/MF brigade training objectives and possibly includes a major field training component. Not all F/MF brigade headquarters need to receive a gold-standard training opportunity for LSCO or to receive gold-standard training every training cycle. For enhanced WFXs, we recommend that the Army *prioritize* specific F/MF brigade headquarters that are most in need of enhanced training opportunities. If these experiments indicate that enhancement provides significant training value that justifies increased costs, the Army could consider implementing these recommendations on a more systematic basis.

Overall, we offer the following seven recommendations:

1. The Army should focus enhanced WFX training opportunities on specific F/MF brigade headquarters with the highest-priority LSCO missions.
2. For prioritized units, the Army should experiment with conducting longer-duration enhanced WFXs, adapted to better meet the training goals of F/MF brigade headquarters.
3. If feasible, the Army should also experiment with including a field training component as part of an enhanced WFX.
4. The Army should provide F/MF brigade headquarters with EXEVALs as provided for in Army training doctrine and should consider whether it is feasible and desirable to associate EXEVALs with enhanced WFXs, for brigades that participate in such exercises.

5. The Army should consider priority F/MF brigade headquarters when sourcing joint and multinational exercises—particularly for brigade types that are not included as WFX training audiences.

6. Army forums should disseminate examples of innovative home station training, and Army guidance should encourage broader implementation—particularly regarding opportunities that could involve permissions and coordination across multiple stakeholders.

7. Certain brigade headquarters lack an organic signal company, such as engineer brigades and expeditionary military intelligence brigades. The Army should study the challenges this causes for home station training and options for mitigation.[4]

---

[4] The lack of an organic signal company is not necessarily a capability gap once the brigade headquarters is deployed for an operation, provided the brigade receives the necessary signal support once in theater.

Acknowledgments

The authors are grateful to the members of U.S. Army Forces Command G-3/5/7, Strategy, Policy, and Transformation Division, who directed and supported this research. We especially want to thank CW5 John Robinson, Kirk Palan, and Kristin Blake for the guidance and support they provided.

We would like to thank the leaders and staff members of the various functional and multifunctional brigade headquarters who took time out of their busy schedules to discuss their experiences and offer their insights. Their support aided this study immensely. For reasons of confidentiality, we do not include the names of specific individuals who participated in unit interviews.

In addition, we would like to thank our peer reviewers, Gian Gentile and Chad Serena of the RAND Corporation and Gen William S. Wallace, U.S. Army (Ret.). We also benefited from feedback given by colleagues at RAND, whose thoughtful comments improved the quality of this report. In particular, we would like to thank Michael Linick. Finally, we would like to thank Mark Hvizda for his help in producing this document.

Abbreviations

AAR	after-action review
ADA	air defense artillery
ADCON	administrative control
ADP	Army doctrine publication
ADRP	Army doctrine reference publication
ARNG	Army National Guard
ASCC	Army service component command
ATP	Army techniques publication
BCT	brigade combat team
BCTP	Battle Command Training Program
BWFX	Brigade Warfighter Exercise
CAB	combat aviation brigade
CASCOM	Combined Arms Support Command
CATS	combined arms training strategies
CoE	Center of Excellence
COIN	counterinsurgency
COMSEC	communications security
CONUS	continental United States
CPX	command post exercise
CSSB	combat sustainment support battalion
CTC	combat training center
DIVARTY	division artillery
E-MIB	expeditionary military intelligence brigade
EN BDE	engineer brigade
ESB	expeditionary signal battalion
ESC	expeditionary sustainment command
EXEVAL	external evaluation
F/MF	functional and multifunctional
FA BDE	field artillery brigade
FM	field manual

FMSU	Financial Management Support Unit
FORSCOM	U. S. Army Forces Command
FTX	field training exercise
HHC	headquarters and headquarters company
HICOM	higher command
HROB	Human Resources Operations Branch
ID	infantry division
JAAT	joint air attack team
JMRC	Joint Multinational Readiness Center
JRTC	Joint Readiness Training Center
LSCO	large-scale combat operations
MCT	Mission Command Training (as part of the WFX event lifecycle)
MCTP	Mission Command Training Program
MEB	maneuver enhancement brigade
MET	Mission Essential Task
METL	Mission Essential Task List
MI	military intelligence
MP	military police
MRE	mission rehearsal exercise
NATO	North Atlantic Treaty Organization
NSFS	naval surface fire support
NTC	National Training Center
OC/T	observer controller/trainer
ODS	Operation Desert Storm
OEF	Operation Enduring Freedom
OIF	Operation Iraqi Freedom
OPCON	operational control
OPLAN	operation plan
OPSGRP	operations group
RC	reserve component
REFORGER	Return of Forces to Germany
SCT	supporting collective task
SOP	standard operating procedure
SUST BDE	sustainment brigade
T&EO	training and evaluation outline
TACON	tactical control
TPFDD	time-phased force deployment data
TRADOC	U.S. Army Training and Doctrine Command

TSC	theater sustainment command
TTSB	theater tactical signal brigade
UAP	unified action partner
USAR	U.S. Army Reserve
WFX	Warfighter Exercise
WSMR	White Sands Missile Range

CHAPTER ONE

Introduction

Mission command involves the way in which "commanders, supported by their staffs, combine the art of command and the science of control to understand situations, make decisions, direct action, and accomplish missions."[1] The goal is to "guide, integrate, and synchronize Army forces throughout the conduct of unified land operations."[2]

The ways that the U.S. Army exercised mission command as part of counterinsurgency (COIN) and stabilization operations during Operation Enduring Freedom (OEF) and Operation Iraqi Freedom (OIF) differs from the ways that the Army would do so as part of large-scale combat operations (LSCO). Because of the Army's focus on COIN and stabilization operations during OIF and OEF, Army leaders have expressed concerns that the abilities of leaders and their staffs to exercise mission command as part of LSCO have atrophied.

The missions conducted in COIN and stability operations as part of OIF and OEF mostly had long lead times for planning, were relatively short in duration, were generally conducted by company-sized units, and generally did not require the detailed planning and synchronization of large formations across space and time. In contrast, LSCO might involve fast-moving, large-scale combined arms maneuvers integrated at the division and corps levels—potentially against a highly lethal adversary. Commanders and staffs must be prepared to track vast amounts of information and to make decisions at a pace dictated by events, rather than by routine planning cycles.

The Army's new operating concept, *The U.S. Army in Multi-Domain Operations 2028*, focuses the Army (as part of the Joint Force) on adapting to the challenges of military conflict with near-peer adversaries.[3] Among other things, this concept addresses the need for rigorous mission command training across echelons. Such training is needed before the start of a conflict, so forces can be ready to deploy and immediately enter the fight in a fast-moving, contested, antiaccess and area denial environment.

> America's adversaries have studied U.S. operations closely. . . . They are deploying capabilities to fight the [United States] through multiple layers of stand-off in all domains—space, cyber, air, sea, and land. . . . In the event of armed conflict, Army forces immediately penetrate enemy anti-access and area denial systems by neutralizing enemy long-range systems. . . . This sets the conditions for a quick transition to joint air-ground operations in which maneuver

---

[1] Army Doctrine Publication (ADP) 6-0, *Mission Command*, Change No. 2, Washington, D.C.: Headquarters, Department of the Army, March 12, 2014, p. ii.

[2] ADP 6-0, 2014, Fig. 1, p. iv.

[3] U.S. Army Training and Doctrine Command (TRADOC), *The U.S. Army in Multi-Domain Operations 2028*, Fort Eustis, Va., TRADOC Pamphlet 525-3-1, November 27, 2018.

enables strike and strike enables maneuver. . . . Army headquarters must not only have the technical, intellectual, and doctrinal tools to execute multi-domain command and control, but rigorous joint and combined training to realize it. . . . The Army must continue to build trusted teams of professionals that thrive in ambiguity and chaos and who are empowered through a doctrine of mission command to rapidly react to threats and opportunities based on a commander's intent. The [multi-domain operation] concept leverages a critical U.S. military advantage—our people. But the Army does not always design our training programs and exercises in ways that facilitate or require this type of decentralized decision making. More intellectual effort is required to improve training designs that facilitate mission command of [multi-domain operation] given the increased complexity.[4]

Purpose

In this report, we examine the effectiveness of mission command training conducted by different types of functional and multifunctional (F/MF) brigade headquarters in preparation for LSCO. That is, we focus specifically on LSCO, rather than addressing the full range of missions that Army F/MF brigade headquarters might be ordered to perform. Our objective is to identify gaps in current training approaches for LSCO and to recommend ways they could be overcome.

Our Approach

At the brigade echelon, the Army is composed of brigade combat teams (BCTs), multifunctional brigades, and functional brigades. Functional brigades provide capabilities specific to a given branch; examples include engineer, military police (MP), air and missile defense, civil affairs, intelligence, and signal brigades. Multifunctional brigades combine the capabilities of several branches; examples include combat aviation, field artillery, sustainment, and maneuver enhancement brigades.

This project's scope did not permit us to examine all types of F/MF brigade headquarters. Instead, in coordination with our U.S. Army Forces Command (FORSCOM) sponsor, we selected the following types of F/MF brigade headquarters as a representative sample:

- combat aviation brigade (CAB)
- field artillery brigade (FA BDE)
- sustainment brigade (SUST BDE)
- engineer brigade (EN BDE)
- expeditionary military intelligence brigade (E-MIB)
- theater tactical signal brigade (TTSB).

Within these types, we focused in particular on training approaches applicable to Regular Army units, as opposed to units in the Army National Guard (ARNG) or U.S. Army Reserve (USAR). Although many of our findings and recommendations are applicable to ARNG and

---

[4] TRADOC, 2018, pp. i, vii, 25, 46, F-3.

USAR units, reserve component (RC) units are also likely to experience circumstances and challenges that differ from those of Regular Army units. Unfortunately, assessing the training approaches most applicable to ARNG and USAR units was simply beyond the scope of this study.

We took a broad view of the scope of mission command training. We not only examined whether training approaches provide F/MF brigade headquarters the opportunity to exercise mission command systems and to practice the steps of the operations process, we also assessed whether approaches allow F/MF brigade headquarters to become trained and proficient in exercising mission command across the range of Mission Essential Tasks (METs) needed for LSCO (with the ability to execute mission command being an element of each such MET). We find this approach is consistent with Army doctrine on mission command, as summarized in Chapter Two.

In assessing mission command training approaches, we looked at how various training approaches and training enablers prepare F/MF brigade headquarters to operate in LSCO, including:

1. overarching doctrine related to mission command, as well as doctrine applicable to specific brigade types
2. training products and enablers, such as Mission Essential Task Lists (METLs), combined arms training strategies (CATS), and training and evaluation outlines (T&EOs)
3. home station training
4. major training opportunities, such as the Warfighter Exercise (WFX) provided by the Mission Command Training Program (MCTP), as well as joint and combined exercises
5. external evaluations (EXEVALs).

Interviews with key stakeholders contributed significantly to our assessment. For reasons of confidentiality, we identify neither the specific individuals we interviewed nor the specific F/MF brigade headquarters. We began by speaking with individuals from MCTP, from the Mission Command Center of Excellence (CoE), and from several other CoEs and schools. (Appendix C contains samples of the questions we prepared for interviews with various stakeholders.) We also observed the June 2017 WFX at Fort Bragg, North Carolina, and spoke with certain participants. These discussions were exceptionally valuable in helping to form our conclusions and recommendations.

We then interviewed leaders and/or key staff members from several F/MF brigade headquarters, including[5]:

- four CAB headquarters
- three SUST BDE headquarters
- two EN BDE headquarters
- one TTSB headquarters
- two E-MIB headquarters.

---

[5] Our goal was to speak with the following individuals in each brigade: the brigade commander, the deputy commander, the executive officer, and the operations officer, as well as any other individuals that unit staff might identify. In some instances, we were only able to speak with a subset of personnel, or our interview was limited to the brigade operations officer.

Unfortunately, we were unsuccessful in our attempts to speak with any FA BDE head-quarters, which limited our insights into this unit type. This shortcoming is at least partially mitigated by the significant, valuable input we received from the Fires CoE, as well as from participants at the June 2017 WFX. In the end, our inability to speak with any FA BDE head-quarters should not fundamentally weaken the overall research effort; the conclusions and recommendation of this report should generally apply to FA BDE headquarters, as well as the other types of F/MF brigades headquarters examined.

We note that our discussion with MCTP occurred relatively early in the project, before we had the opportunity to speak with the various units we interviewed. Many of our recom-mendations relate to ways in which the Army might adapt the WFX to better address the training objectives of at least some participating brigade headquarters. Unfortunately, we were unable to meet with MCTP for a second time at the end of the project to get its perspective on our recommendations. If the Army chooses to move ahead with implementing any of our recommendations that relate to MCTP, it should ensure that MCTP has an opportunity to review and provide input before implementation.

Finally, we stress that there is a distinction between the training proficiency of an F/MF *brigade headquarters* in its own right and the training proficiency of an entire F/MF *brigade* as a multiechelon force package—i.e., from the brigade headquarters down through subordinate battalions and companies. This study focuses on whether current training approaches allow F/MF *brigade headquarters* to become trained and proficient in exercising mission command across the range of METs required of LSCO. This study *does not* assess training approaches for the various types of battalions and companies that operate subordinate to different F/MF bri-gade headquarters. Any conclusions that we might draw regarding gaps or challenges in train-ing approaches for F/MF brigade headquarters are not intended to imply that comparable gaps and challenges necessarily apply at subordinate echelons.

Organization of This Report

The remainder of the report proceeds as follows. Chapter Two first provides an overview of Army doctrine and training products (e.g., METLs, CATS, T&EOs) pertinent to mission command, summarizing related insights derived from stakeholder interviews. (Appendix A describes the doctrinal organization, missions, and command and support relationships in more detail for the types of F/MF brigade headquarters we looked at; it includes METLs for those brigade types as reference.) The chapter then includes descriptions of current home sta-tion training approaches for F/MF brigade headquarters. Chapter Three begins by summa-rizing opportunities that existed prior to OIF for F/MF brigade headquarters to participate in large-scale training and exercise opportunities. This provides background and context for thinking about similar opportunities available at present. (Appendix B provides a more detailed discussion of this topic.) In Chapter Three, we describe current larger-scale training opportuni-ties—emphasizing MCTP's WFX—and discusses how EXEVALs factor into F/MF brigade training approaches. Many of our recommendations relate to how the Army could adapt the WFX to further enhance the training of F/MF brigade headquarters for LSCO. Chapter Four is a summary of our major findings and recommendations. Appendix C contains samples of the questions we prepared for interviews with various stakeholders.

CHAPTER TWO

Doctrine, Training Products, and Home Station Training

This chapter begins with an overview of Army doctrine and training products pertinent to mission command, including insights on their utility derived from stakeholder interviews. We then describe current home station training approaches for F/MF brigade headquarters. We identify examples of innovative approaches that could be used as models for other units. We also discuss an issue that certain brigade types highlighted as a significant challenge for home station training: the lack of an organic signal company. Throughout this chapter, we provide recommendations for improvement, based on our research.

Doctrine and Training Products Pertinent to Mission Command

In this section, we provide an overview of Army doctrine and training products pertinent to mission command, including insights on their utility derived from stakeholder interviews. We begin by reviewing foundational, Army-wide doctrine for mission command. Next, we discuss other branch-specific doctrine that describes the missions and capabilities for specific brigade types. Then, we review training products that the Army develops to help translate doctrine into more-structured guidance and enablers to support unit training. Key products include METLs, T&EOs, and CATS. Finally, we summarize insights derived from interviews with units and CoEs related to these topics. Appendix A provides additional information on doctrine describing how certain types of F/MF brigade headquarters are expected to operate in LSCO. It also includes METLs for those brigade types as reference.

Foundational Doctrine

ADP 6-0, *Mission Command*, presents the Army's foundational mission command guidance. Along with ADP 5-0, *The Operations Process*, ADP 6-0 forms the foundation for the tactics, techniques, and procedures for the exercise of mission command.[1] ADPs 5-0 and 6-0 each have an associated, more detailed ADRP. Other important doctrine includes Field Manual (FM) 6-0, *Commander and Staff Organization and Operations*, and FM 7-0, *Train to Win in a Complex World*.[2] The U.S. Army Combined Arms Center is a proponent for all these documents.

---

[1] See ADP 6-0, 2014; ADP 5-0, *The Operations Process*, Washington, D.C.: Headquarters, Department of the Army, May 17, 2012; and Army Doctrine Reference Publication (ADRP) 5-0, *The Operations Process*, Washington, D.C.: Headquarters, Department of the Army, May 17, 2012.

[2] FM 6-0, *Commander and Staff Organization and Operations*, Washington, D.C.: Headquarters, Department of the Army, May 11, 2015; and FM 7-0, *Train to Win in a Complex World*, Washington, D.C.: Headquarters, Department of the Army, October 5, 2016.

Mission command involves the way in which "commanders, supported by their staffs, combine the art of command and the science of control to understand situations, make decisions, direct action, and accomplish missions."[3] The goal is to "guide, integrate, and synchronize Army forces throughout the conduct of unified land operations."[4] Per ADRP 6-0, "mission command emphasizes centralized intent and dispersed execution through disciplined initiative."[5] In other words, the goal of mission command doctrine is to enable subordinate units to execute their assigned missions in ways that they deem most appropriate—as long as their actions are consistent with the intentions of higher commanders regarding the purpose of the mission, the key tasks, the desired end state, and the available resources. Military operations are inherently complex and uncertain, characterized by fog and friction, and therefore subordinates must be able to adapt to situations without seeking immediate guidance from higher commanders and seize and exploit the initiative more rapidly than the adversary.

The *operations process* is the process by which commanders and staffs plan, prepare, execute, and assess military operations.[6] As ADP 6-0 states, "commanders use the operations process as the overarching framework for exercising mission command."[7] Exercising mission command through the operations process involves both a foundational *philosophy* and a *warfighting function*. The mission command *philosophy* articulates a set of principles that guide the *spirit* by which commanders and staffs execute the operations process:

- Build cohesive teams through mutual trust.
- Create shared understanding.
- Provide a clear commander's intent.
- Exercise disciplined initiative.
- Use mission orders.
- Accept prudent risk.[8]

The mission command *warfighting function* focuses on the mechanics of the various *tasks* that commanders and staffs must perform to conduct the operations process. These *tasks* are enabled by various types of mission command *systems* (e.g., networks and information systems).

Mission command is about much more than simply how a commander and staff perform the operations process internal to their headquarters to develop mission orders that reflect the commander's intent. As a commander leads the operations process at a given echelon, they must remain cognizant of the commander's intent up to two echelons above them—including specified and implied tasks—to achieve unity of effort.[9]

---

[3]   ADRP 6-0, Mission Command , Change No. 2, Washington, D.C.: Headquarters, Department of the Army, March 28, 2014, p. ii.

[4]   ADRP 6-0, 2014, Fig. 1.

[5]   ADRP 6-0, 2014, p. 1-1.

[6]   The operations process involves "the major mission command activities performed during operations: planning, preparing, executing, and continuously assessing the operation . . . . Commanders, supported by their staffs, use the operations process to drive the conceptual and detailed planning necessary to understand, visualize, and describe their operational environment; make and articulate decisions; and direct, lead, and assess military operations" (ADRP 5-0, 2012, p. 1-2).

[7]   ADP 6-0, 2014, p. 9.

[8]   ADP 6-0, 2014, p. 2.

[9]   ADRP 5-0, 2012, p. 2-18; ADP 6-0, 2014, p. 4.

Likewise, mission command is about more than just how a headquarters at a given echelon interacts with its subordinate elements. Communications and synchronization must occur across echelons—higher, parallel, and subordinate—and might also involve synchronization with unified action partners (UAPs). UAPs might include joint and other nations' military forces, governmental and nongovernmental organizations, and private-sector organizations. As ADRP 6-0 states:

> Through mission command, commanders integrate and synchronize operations. Commanders understand they do not operate independently but as part of a larger force. They integrate and synchronize their actions with the rest of the force to achieve the overall objective of the operation. . . . Effective staffs establish and maintain a high degree of coordination and cooperation with staffs of higher, lower, supporting, supported, and adjacent units. They do this by actively collaborating with commanders and staffs of other units to solve problems. . . . The traditional view of communication within military organizations is that subordinates send commanders information, and commanders provide subordinates with decisions and instructions. This linear form of communication is inadequate for mission command. Communication has an importance far beyond exchanging information. Commanders and staffs communicate to learn, exchange ideas, and create and sustain shared understanding. Information needs to flow up and down the chain of command as well as laterally to adjacent units and organizations. Separate from the quality or meaning of information exchanged, communication strengthens bonds within a command. It is an important factor in building trust, cooperation, cohesion, and mutual understanding.[10]

Finally, proficiency in mission command is not an independent objective, per se. Proficiency in mission command enables the proficient execution of tasks and functions to accomplish specific missions. According to ADRP 6-0:

> The mission command warfighting function integrates the other warfighting functions (movement and maneuver, intelligence, fires, sustainment, and protection) into a coherent whole. By itself, the mission command warfighting function will not secure an objective, move a friendly force, or restore an essential service to a population. Instead, it provides purpose and direction to the other warfighting functions. Commanders use the mission command warfighting function to help achieve objectives and accomplish missions.[11]

As we assessed mission command training approaches, we looked at whether these approaches allow F/MF brigade headquarters to become trained and proficient in exercising mission command across the range of wartime tasks.

Doctrine and Organization Particular to Specific Types of F/MF Brigade Headquarters
In addition to the foundational doctrine described above that applies across the Army, other branch-specific doctrine describes the missions and capabilities for specific brigade types.[12] The respective branch CoEs and schools are proponents for this doctrine. Branch-specific doc-

---

[10] ADRP 6-0, 2014, p. 2-14.

[11] ADRP 6-0, 2014, p. 1-4.

[12] Branch-specific doctrine generally summarizes the key elements of ADRP 6-0 and other foundational doctrine, at least as an overview.

trine sets the context for how different brigade types exercise mission command. As one CoE interviewee noted, the basic process for exercising mission command—i.e., the operations process—is the same for every brigade type in the Army; however, the specifics of managing mission execution vary based on the specific mission and tasks of each brigade type.

For many brigade types, there are multiple branch-specific doctrinal publications providing different levels of technical detail—for example, a less detailed FM that provides branch-wide doctrine and a more detailed Army techniques publication (ATP) focused on a specific unit type. There might also be relevant doctrine in an ADP and an ADRP. For example, branch-specific doctrine relevant to FA BDEs includes the following publications:

- ADRP 3-09, *Fires*
- FM 3-09, *Field Artillery Operations and Fire Support*
- ATP 3-09.24, *Techniques for the Fires Brigade*.[13]

Various publications are updated at different points in time, so keeping the branch-specific doctrine synchronized across publications—and keeping these synchronized with foundational mission command doctrine—involves a lag time that can be considerable. In short, there is not a single publication that a unit can go to for all relevant doctrine—not an insurmountable challenge, but one that can make the achievement of doctrinal fluency somewhat burdensome. Moreover, as the Army adds new unit types, such as division artillery (DIVARTY) headquarters and E-MIBs, there is a time lag in developing unit-specific doctrine and integrating it into branchwide doctrine.

A key principle of mission command philosophy is to "build cohesive teams through mutual trust."[14] Mission command doctrine states that team-building is based on trust, shared experiences, and training, and it takes time to develop. BCTs, for example, have an organic set of subordinate battalions they train with and (in most cases) deploy with, maximizing the benefits of long-term habitual relationships for mission command. (See Appendix A for a summary of Army command and support relationships.) F/MF brigades generally do not have a full set of organic subordinate units that they will deploy and operate with in LSCO, which constrains their ability to build cohesive teams through habitual relationships.

CABs are the major exception. CABs have six organic subordinate elements—four rotary-wing battalions, a Gray Eagle unmanned aerial system company, and an aviation support battalion for sustainment—that normally deploy and operate with the CAB headquarters. Some types of F/MF brigades, such as FA BDEs and SUST BDEs, have a limited set of organic subordinates that provide signal or sustainment functions, but during operations, all other subordinates are attached based on mission requirements. Other types of F/MF brigades, such as EN BDEs and TTSBs, have no organic subordinates, meaning that all subordinates are attached based on mission requirements. For all of these brigade types, a given brigade headquarters might deploy with subordinate elements that are under its habitual administrative control (ADCON) at home station; on the other hand, the brigade's deployed task organization might

---

[13] ADRP 3-09, *Fires, Change No. 1*, Washington, D.C.: Headquarters, Department of the Army, February 8, 2013; FM 3-09, *Field Artillery Operations and Fire Support*, Washington, D.C.: Headquarters, Department of the Army, April 4, 2014; ATP 3-09.24, *Techniques for the Fires Brigade*, Washington, D.C.: Headquarters, Department of the Army, November 21, 2012.

[14] ADP 6-0, 2014, p. 2.

include only attached subordinates with no habitual relationships. (The E-MIB is a special case. It has two organic battalions to deploy with, but in theater these might be attached to other commands, rather than remaining under E-MIB headquarters' direct command.)

Moreover, building cohesive teams through mutual trust is not simply an issue between the brigade headquarters and its subordinate elements. For brigade headquarters, for example, team building also involves integrating with higher and parallel echelons. Higher echelons include division and corps headquarters; parallel echelons include other types of brigades that a given F/MF brigade headquarters must synchronize with to perform its functions. For example, a CAB is normally attached to a division headquarters and must synchronize its functions with BCTs, the FA BDEs, DIVARTY, and (to a lesser extent) the air defense artillery (ADA) brigade. In addition, the CAB's subordinate elements must often integrate with BCTs or other supported formations. Although the CAB's subordinate elements will generally remain under the command of the CAB headquarters and exercise support relationships with other formations, the CAB's subordinate elements could be placed under the operational or tactical control (TACON) of other commands for some missions. CAB doctrine emphasizes that habitual training with the supported maneuver element decreases risk and aids mission performance:

> The ground maneuver commander in close enemy contact controls the synchronization and integration of Army Aviation maneuver and the distribution and deconfliction of Army Aviation fires. Shared understanding within the combined arms team, through known standardized procedures and habitual training, increases the likelihood of successful employment of attack reconnaissance units against enemy forces in close contact with friendly forces. However, during in extremis situations, Army Aviation attack reconnaissance units may conduct hasty attacks in support of all friendly ground units regardless of their training level or habitual relationship, but with greater risk.[15]

The key point here is that, in LSCO, most types of F/MF brigade headquarters *by design* will likely not exercise mission command over habitually aligned subordinates.[16] Moreover, even for CABs and BCTs, there is no certainty that a given brigade will deploy alongside the parallel and higher echelons with which it has habitual relationships at home station. In LSCO, for example, a CAB could be attached to a different division headquarters. It could need to support BCTs from divisions that it has never trained with. It might need to integrate with an FA BDE that it has likewise never trained with.

In short, although mission command doctrine stresses the desirability of long-term team-building through shared relationships and training, in an actual LSCO, rapid team-building in theater will probably be necessary across echelons. Even brigade headquarters with organic subordinates might still face this challenge with regard to parallel and higher echelons. For now, we simply describe the challenge. In the next chapter, we discuss how units characterized the attendant risks and offer recommendations that could mitigate challenges (at least in some circumstances for some units).

---

[15] FM 3-04, *Army Aviation*, Washington, D.C.: Department of the Army, July 29, 2015, pp. 3-3–3-4.

[16] Habitual alignments could be organic relationships, as with BCTs and CABs, or based on enduring peacetime ADCON relationships, as exist for other types of F/MF brigade headquarters.

As reference, Table 2.1 summarizes various organization relationships for the types of F/MF brigade headquarters that we focus on in this study. (Appendix A describes doctrinal organization, missions, and command and support relationships in more detail.)

Training Products

The Army develops a range of training products that help translate doctrine into more-structured guidance and enablers that support unit training. Key products include METLs, T&EOs, and CATS. Training products originate from two different sources. For a given unit type, most elements originate with the branch proponent. However, the Mission Command CoE develops products with a mission-command focus that apply Army-wide.

For example, branch proponents develop METLs. METLs are made up of several individual METs. A MET provides a doctrinal framework of fundamental collective tasks to perform the unit's wartime mission. Each MET has several supporting collective tasks (SCTs). METLs for brigade headquarters do not include a specific MET for mission command, but "Conduct Mission Command Operations Process" is an SCT to each MET on each brigade

Table 2.1
Organizational Relationships for Certain F/MF Brigade Headquarters

Brigade Type	Organic Subordinates That Deploy and Operate with Brigade Headquarters	Typical Higher Command for Brigade Headquarters in LSCO	Other Key Units for Integration in LSCO (brigade and above)
CAB	Yes; six organic subordinate battalions/companies normally deploy and operate with CAB headquarters	Division headquarters	BCT, FA BDE/DIVARTY, and ADA brigade
FA BDE	Limited; brigade support battalion headquarters and signal company are organic, but other subordinate elements are task organized	Corps headquarters	Division headquarters/ DIVARTY, BCT, CAB, ADA brigade
SUST BDE	Limited; signal company is organic, but other subordinate elements are task organized	ESC	Customers for whom the SUST BDE provides sustainment support; "ground owning" brigades (e.g., BCTs, MEBs, MP brigades) that manage transit through an area of operations and/or provide security for the SUST BDE and its subordinates (if applicable)[a]
EN BDE	No; all subordinate elements are task organized	Corps headquarters	Division headquarters, BCT, MEB
TTSB	No; all subordinate elements are task organized	Corps headquarters	Supported customers (units without organic signal)
E-MIB	Partial; two organic battalions that might deploy but, if deployed, might be attached to other commands; brigade headquarters might also command other nonorganic subordinates	Corps headquarters	Division headquarters, BCT, CAB (to coordinate Gray Eagle missions)

SOURCE: See Appendix A for more information on these relationships.
NOTE: ESC = expeditionary sustainment command; MEB = maneuver enhancement brigade.
[a] For example: "The maneuver enhancement brigade may be responsible for the terrain assignment and establishing secure movement corridors. The sustainment brigade base will be integrated into area terrain management and protection plans will be based on established command and support relationships and the physical space occupied. Within the support area, the sustainment brigade answers to the maneuver enhancement brigade, or the identified terrain owner, for protection, security, and related matters" (ATP 4-93, *Sustainment Brigade*, Washington, D.C.: Headquarters, Department of the Army, April 11, 2016, p. 5-5).

headquarters' METL. The Mission Command CoE is the proponent for the mission command SCT. Figure 2.1 provides an example, using the METL for an EN BDE headquarters. METs are shown in blue. SCTs are only shown for a single MET: Conduct Survivability Operations (brigade). For this MET, the mission command SCT is shown in red. (Appendix A includes METLs for the brigade types we focus on in this study.)

Again, METLs represent a doctrinal framework for training. T&EOs and CATS are key products that guide units in how to train to METL standards. There are T&EOs for each MET and for each SCT. Each T&EO outlines major process steps and standards of performance. The branch proponent responsible for each MET or SCT develops and approves its T&EOs. Using Figure 2.1 as an example, the Engineer School develops T&EOs for each MET and for each SCT other than "71-Brigade-5100—Conduct Mission Command Operations Process for Brigades." The Mission Command CoE develops the T&EO for that SCT.

Units are supposed to focus on METs when evaluating training readiness; MET T&EOs are supposed to form the basis for evaluation. SCTs are intended to help inform MET assessments, but they are not assessed individually. In other words, units are not required to explicitly evaluate each SCT, only the overall MET (which has its own T&EO). For example, the T&EO for the mission command SCT is intended to be used in conjunction with T&EO for the overall MET, to ensure the unit is completing the operations process in a manner consistent with mission command doctrine. Again, however, the unit is not explicitly evaluated on mission command as a stand-alone proficiency. Evaluations are based on the T&EO for the MET, with the T&EO for the mission command SCT providing context.[17] According to FM 7-0, units are supposed to receive an EXEVAL led by commanders one and two echelons up to be validated as T (fully trained) or T- (trained) on a given MET. We return to the issue of EXEVALs at the end of Chapter Three.

CATS provide a baseline strategy to guide units as they plan, prepare, execute, and assess training. CATS include different types of events (staff exercises, command post exercises [CPXs], field training exercises [FTXs], etc.) and identify ways to train the tasks contained in each MET. CATS offer high-level guidance on how mission command factors in as an element of the purpose, execution guidance, and desired outcome for each training event (where applicable). In any case, there is no specific requirement that units use any of the CATS products

Figure 2.1
METL for an Engineer Brigade Headquarters

05-Brigade-0096 - Conduct General Engineering Operations
05-Brigade-0098 - Conduct Countermobility Operations
05-Brigade-0099 - Conduct Survivability Operations
     05-Brigade-0080 - Recommend Employment Priority of Engineer Assets
     05-Brigade-0812 - Conduct Equipment Support Missions
     71-Brigade-5100 - Conduct Mission Command Operations Process for Brigades
05-Brigade-1007 - Conduct Mobility Operations
55-Brigade-4800 - Conduct Expeditionary Deployment Operations at Brigade Level

SOURCE: Army Training Network, *METL Viewer*, undated.

---

[17] U.S. Army Combined Arms Center, "FM 7-0 Tutorial: Overview and How to Use the Training & Evaluation Outline (T&EO)," briefing slides, March 22, 2017.

to develop training strategies. CATS are simply a training aid to help units develop their own training strategies.

Insights from Stakeholder Interviews
Mission command training is governed by a set of related and, in some cases, overlapping doctrinal publications. Units we spoke with had mixed opinions about both the utility of the documents that existed as resources and the value of having so many different types of documents. Some units felt the quality of products is good overall, as is the training on how to use them. Others noted areas for improvement. There were few unifying themes from the responses, however, and few concrete recommendations as to how doctrine, training products, or courses of instruction should be modified to improve training at the brigade level. Most seemed to feel that doctrine and training products were good enough, and most training was done on the job. To paraphrase one interviewee: Doctrine and training products are generally sufficient; professional military education is good if soldiers do it; if a unit fails, it is not the schooling but the lack of practice. With regard to on the job training, the consensus seemed to be that, by the time key staff members were assigned to brigade-level positions, they were functionally proficient with doctrine and training products, and then continued to learn as needed as they performed their jobs. However, several noted that challenges were bigger at lower echelons for soldiers who were at earlier stages of their careers. A final theme we heard several times from interviewees at the CoEs was the worry that the proliferation of different products containing information relevant to mission command could be confusing to those consuming the doctrine. One interviewee stated that there used to be just one FM, but now doctrine is spread across various ADPs, ADRPs, and other publications, which are updated at different points in time. This interviewee said that it would be easier if foundational doctrine was consolidated in one place.

Because we received no consensus input from stakeholders as to how changes to doctrine and training products could significantly enhance the training of F/MF brigade headquarters, we make no specific recommendations on this topic.

Home Station Training

In this section, we describe home station training approaches for F/MF brigade headquarters. We identify examples of innovative approaches used by units that we interviewed; these could be used as models for other units to replicate. We also discuss an issue that certain brigade types highlighted as a significant challenge for home station training: the lack of an organic signal company.

Overview
Home station is an important venue for mission command training, but there are limitations. To paraphrase one interviewee from the Mission Command CoE: Most mission command training occurs at the home station, especially for the F/MF brigade headquarters that do not participate as WFX training audiences (a topic we discuss in the next chapter). Although such training can be effective, it can only be so comprehensive. The major home station training events are generally CPXs and FTXs. The units we interviewed generally reported that in normal years—i.e., years absent a deployment or preparation for a WFX—they attempted to

execute some combination of CPXs and FTXs two to four times a year. Some events focus on operations internal to the brigade headquarters. Other events might involve multiechelon training with subordinates. No brigade headquarters that we spoke with identified home station training opportunities that included a higher command (HICOM) providing exercise direction or opportunities to integrate with parallel and higher headquarters as part of a home station training exercise.

Some brigade headquarters noted challenges in conducting FTXs with subordinates. CABs were less challenged in this regard, as they are co-located with an organic set of subordinate battalions that normally deploy with the brigade headquarters, making it comparatively easy to synchronize training and deployment cycles—even allowing for the fact that aviation battalions remain very busy supporting BCT training and performing other nondeployed missions.

Other brigade types noted greater challenges, particularly if subordinates were not co-located or were frequently on their own deployment cycles. For example, staff for one EN BDE we interviewed noted that it has ADCON over subordinate battalions at four installations; when factoring in subordinates below battalion level, the count increases to seven installations. This brigade attempts to do one FTX per quarter, but the number of subordinates that can participate is limited. In any case, these subordinates are only administratively aligned, not organic; they might not be the same subordinates that the brigade headquarters would command during an LSCO. When the brigade holds FTXs, these generally focus on construction (i.e., the Plan General Engineer Operations MET); other METs are less conducive to home station training. A second EN BDE we spoke with noted that they attempted to conduct some CPXs with participation from subordinates, but it was rare to have an FTX with subordinates.

Frequent subordinate deployments or home station taskings also compete with time available for home station training. One SUST BDE emphasized how challenging it can be to make time for dedicated training events, given the pressure of other taskings (comments are paraphrased rather than a direct quote):

> Time is a significant challenge. We support up to three combat training center (CTC) events per year. There are many last-minute taskings and frequent deployments. The brigade can plan training, but it is hard to execute. We are "living in amber" from a readiness perspective. We never get to a high level because of constant personnel turnover.

The E-MIBs and TTSBs we spoke with noted similar challenges. According to the TTSB personnel we interviewed (comments are paraphrased rather than a direct quote):

> Most expeditionary signal battalions (ESBs) are not on a typical train-deploy-reset schedule. Companies and detachments deploy all the time. It is a more or less constant reality to have a portion of the TTSB deployed. This is a significant burden on home station training. When we do brigade FTXs, we do some of it with a cross-leveled pick-up team, meaning that, while the FTXs are useful, some personnel are operating in positions for which they do not have the technical qualifications.

The E-MIBs we spoke with also noted that they are on a high-frequency deployment schedule. One noted that most of their training was deployment-focused, rather than geared to LSCO preparation. Another E-MIB explained that (comments are paraphrased rather than a direct quote):

The strain of deployments is an issue. The brigade frequently deploys small detachments. When it does so, it attempts to send additional leaders as well to provide leadership support for the detachment, but this negatively impacts operations at home. It is hard to grow enough [military intelligence (MI)] structure to meet demand.

Several of the brigade headquarters we spoke with noted that they had challenges training all METs at home station, or that they had trouble setting the conditions at home station to adequately support their training objectives. Above we mentioned the EN BDE that stated most home station training is focused on the General Engineer Operations MET. One SUST BDE stated that it gets opportunities to train most METs. However, METs related to deployment and redeployment operations could be challenging, although providing day-to-day customer support helps. This SUST BDE found that the MET "Conduct Distribution Operations" was the most difficult to train at home station training, as exercising this MET involved the need to manage multiple subordinate battalions. The TTSB we spoke with stated that its biggest challenge is staging exercises for which they actually have customers to support; without this, the exercises are less useful. They also do not have as many opportunities to practice under degraded conditions. (An interviewee at the Cyber CoE likewise noted that mission command systems are hard to use, and it is hard to learn to use them when a unit does not have any customers to serve as practice.) One E-MIB noted that its subordinate elements frequently support other training audiences at dirt CTCs[18] and elsewhere; although this is valuable up to a point, the nature of CTC events are understandably focused on the maneuver force and are not structured to directly support the training objectives of E-MIB elements. When the E-MIB attempts to do a home station event of its own, however, it is unable to successfully request maneuver support (even at platoon level) to support its training objectives.

Divisional F/MF brigade headquarters noted that providing support to BCTs preparing for CTC rotations added at least some training value and experience. Most BCT support comes from elements subordinate to the F/MF brigade headquarters, rather than from the brigade headquarters itself. Still, some brigade headquarters noted that they attempt to make a mission command exercise out of relaying orders from the division headquarters to the brigade's subordinate elements operating in support of the BCT. Multiple brigade headquarters noted that it was sometimes a challenge to act in this role, as the division headquarters often attempted to relay orders directly to subordinate battalions, rather than through the brigade headquarters. Some brigade headquarters identified other innovative ways they train in relation to a BCT's CTC rotation, which we discuss later in this report.

Examples of Innovative Training Approaches

Some brigade headquarters report innovative ways of exercising subordinates as part of home station training. For example, one SUST BDE noted that when it sends subordinate elements to support BCT rotations at a CTC, it tries to send elements of the brigade headquarters to embed with the BCT and exercise mission command over the SUST BDE's subordinates. The embedded elements can practice the concept of support, how to track supplies, and how to communicate up and down the chain of command (up to the G4 and down to the combat sustainment support battalion [CSSB]). This helps to give a foundation in exercising mis-

---

[18]   The National Training Center (NTC), the Joint Readiness Training Center (JRTC), and the Joint Multinational Readiness Center (JMRC). The JMRC was then named the Combat Maneuver Training Center.

sion command. So far, the CTC has been supportive, but it also requires the approval of the brigade's division headquarters and the BCT training audience. One key is to show that the SUST BDE's embedded element is taking a burden off the training audience and the CTC white cell, rather than simply furthering its own training objectives.

As noted above, one EN BDE staff member that we spoke with has administrative relationships with subordinate units across seven installations. When it conducts FTXs (which are focused on construction), it tries to synchronize these events with projects that the installations want done to add training value. The relationships that support these exercises are based on the personalities of the various commanders involved.

One E-MIB noted that when different elements are providing support to other training audiences simultaneously, the brigade attempts to link them in ways that help practice aspects of mission command. For example, they noted an instance where elements of the brigade headquarters embedded with a corps training audience at a WFX to support the corps G2 in executing its intelligence, surveillance, and reconnaissance function. One subordinate battalion also supported the WFX as a response cell and had teams supporting a BCT's CTC rotation. The other subordinate battalion was at an exercise practicing signals intelligence collection and interrogation with an MP company. The E-MIB noted that "stacking" events like this provides time and flexibility for retraining.

Recommendation One
Across brigade types, encourage and disseminate examples of innovative home station training. We recommend that the Army establish a forum to collect and disseminate innovative training examples as part of the lessons learned process. Dissemination though the Center for Army Lessons Learned could serve as one method, although ensuring the systematic collection and broad consumption of disseminated products across the community of F/MF brigade headquarters could be a challenge. A second, and perhaps more effective, method could involve establishing a semiformal community of practice/interest among F/MF brigade headquarters, under FORSCOM's chairmanship to encourage active participation across member brigade headquarters. The Army should also consider addressing innovative training examples as part of FORSCOM annual training guidance or a similar forum; this could help stimulate implementation and coordination among multiple stakeholders.

Lack of an Organic Signal Company for Certain Types of Functional Brigade Headquarters
Some brigade headquarters we interviewed identified the lack of an organic signal company as a capability gap that presents a critical challenge for home station training. (In other words, the lack of an organic signal company is not necessarily a capability gap once the brigade headquarters is deployed for an operation, provided the brigade receives the necessary signal support in theater as specified in doctrine.) We interviewed two E-MIB and two EN BDE headquarters. Both unit types lack an organic signal company, and all four units we interviewed highlighted this capability gap as a critical challenge for training. One EN BDE characterized the issue as follows (comments are paraphrased rather than a direct quote):

> One big issue with mission command systems is that we cannot exercise them without access to a tactical network, but that capability resides in another element and we only get access when deployed or for major training. We have no organic access to a Command Post Node (CPN) or Joint Network Node (JNN). We must rely on an ESB. We cannot

have mission command systems online during day-to-day activities. We only get to exercise them at a WFX or a major home station FTX. This amounts to only a few weeks per year.

In addition, one SUST BDE headquarters *with an organic signal company* independently noted its value. The benefit was that the brigade could use its tactical network on a regular basis ("without that, we would struggle"), and could also provide signal support to subordinate elements without having to coordinate with higher headquarters.

On the other hand, an interviewee from the TTSB stated that ESBs have the capacity to support home station training for these brigade headquarters. The interviewee wondered what events were not being supported, and suggested it is simply an issue of planning—that brigade headquarters were deficient in making requests in advance of training where they needed signal support. The interviewee suggested that brigades would find it burdensome to take on the requirement to train and sustain an organic signal company, and that ESBs should be able to handle perhaps double their current taskings.

Another interviewee from an EN BDE responded that ESBs want to lock in training requests 180 days out from a supported training event. This interviewee noted that the brigade headquarters or its subordinate elements would benefit from signal support at many smaller exercises, not just major FTXs; for these, however, it was generally not possible to lock in training requests so far out. This interviewee also asserted that the brigade had made past requests for signal support that were denied, and that ESBs were on their own training paths that might not synch with the needs of the requesting brigade.

In short, the issue appears to be that ESBs can support major training—perhaps at higher rates than they are currently tasked—but brigade headquarters without organic signals want the ability to use tactical networks on a regular basis without significant advance planning.

Recommendation Two
For certain types of brigade headquarters that lack an organic signal company, the Army should study the challenges this causes for home station training, as well as options for mitigation. For brigade headquarters that currently lack an organic signal company, we recommend that the Army assess the costs and benefits of providing a sustained signal capability to support home station training. Unfortunately, it was beyond the scope of this study to assess the issue at a level of detail sufficient to offer a more concrete recommendation. However, the issue appears to have enough merit to warrant further study.

It is possible that the cost-benefit calculus is more in favor of some brigade types than others. Moreover, for a certain brigade type, the cost-benefit calculus might only warrant that this capability be provided to specific units—e.g., high-priority units identified for sourcing against a combatant command's operation plan (OPLAN). We discuss this policy in more detail in the next chapter. In such cases, the answer might not be an organic signal company. For example, it might be feasible to attach an element organic to an ESB to support brigade training on an extended, but not permanent, basis (particularly if that element is also identified for OPLAN sourcing and might even be employed in theater to support the same brigade headquarters).

CHAPTER THREE

Large-Scale Training and Exercises, Including MCTP's WFX

This chapter focuses on large-scale training opportunities for F/MF brigade headquarters. First, we summarize opportunities that existed before OIF for F/MF brigade headquarters to participate in large-scale training and exercise opportunities. This provides background and context for thinking about similar opportunities available now. Next, we discuss the primary larger-scale opportunity currently available—MCTP's WFX. Many of our most significant recommendations relate to how the Army could adapt the WFX to further enhance the training of F/MF brigade headquarters for LSCO. We also touch on specific training opportunities related to current joint and combined exercises that are relevant to the discussion. Finally, we discuss how EXEVALs factor into F/MF brigade training approaches.

Large-Scale Training and Exercise Opportunities Before OIF

The section describes opportunities for participation in major training events and exercises that were available to at least certain types of supporting brigades (e.g., aviation, field artillery, engineer) from the latter portion of the Cold War through the start of OIF. (We refer to *supporting brigades* rather than *F/MF brigades*, since the latter term is a product of Army modularity that occurred during the 2000s and was not in use prior to OIF.[1]) We first discuss the Army's CTC program. We next discuss large-scale joint and combined exercises. Finally, we discuss training opportunities prior to the start of LSCO in Operation Desert Storm (ODS) and OIF. Appendix B provides a more detailed discussion of this topic.

The Army's CTC Program

Today's MCTP provides a more substantial training opportunity for participating F/MF brigade headquarters than existed prior to OIF. As it does today, the Army's CTC program prior to OIF included the three dirt CTCs that focused on maneuver brigades, as well as the Battle Command Training Program (BCTP), which has since been renamed as MCTP.[2] Unlike

---

[1] *Modularity* was a major force redesign initiative that the Army began to implement in 2004 (although its roots run at least into the 1990s in terms of planning, concepts, and initial force redesign efforts). At its core, modularity transformed the Army from a division-based to a brigade-based force—i.e., one focused on BCTs and various types of F/MF brigades that could deploy independently, and one that could be more easily tailored into force packages to meet combatant commander requirements that did not necessarily require a full division with all its organic elements.

[2] Although not formally a part of today's CTC program, the Joint Pacific Multinational Readiness Capability is "a deployable, professionally trained [operations group (OPSGRP)] capable of providing standardized, instrumented EXEVALs to build, maintain, and extend BCT readiness." It provides BCT commanders in the Pacific area of operations with "a CTC 'like' experience with academy-trained OC/T, [an opposing force], and fully instrumented [after action reviews]." See 196th Infantry Brigade, "Joint Pacific Multinational Readiness Capability," webpage, undated.

today's MCTP, however, the BCTP prior to OIF did not include supporting brigade headquarters as a *formal* training audience, even on a secondary basis. Supporting brigade headquarters still received *informal* training value through their participation in support of a division or corps headquarters. Moreover, prior to modularity, some types of brigade headquarters operated with the brigade commander, and substantial portions of the staff embedded with the headquarters of the division or corps to which they were aligned, meaning that BCTP provided a significant experience in terms of integrating with these higher echelons. On the other hand, WFXs prior to OIF were not designed to meet the training objectives of supporting brigade headquarters, and they received no dedicated support from BCTP observer controllers/trainers (OC/Ts). BCTP first began including certain types of F/MF brigade headquarters as a training audience in 2007.[3] Although today's MCTP is an improvement over the pre-OIF BCTP, at least some supporting brigade headquarters received other large-scale training opportunities to help prepare for LSCO.

Large-Scale Joint and Combined Exercises

Large-scale joint and combined exercises provided such opportunities, for at least *some* supporting brigade headquarters. Such exercises often gave participating brigade headquarters a chance to command substantial subordinate formations as part of LSCO-focused field training. REFORGER (Return of Forces to Germany) in Europe and TEAM SPIRIT in South Korea were the best examples. These were generally corps-level exercises, often including maneuver of division-sized elements. These large-scale exercises provided participating brigade headquarters an opportunity to train and conduct mission command in a complex field environment that allowed movement and maneuver of substantial subordinate formations and required coordination and integration as part of a division- or corps-level combined arms team.

In the United States, the BORDER STAR and GALLANT EAGLE series were examples of large-scale joint exercises in the U.S. desert southwest. Such exercises generally included events distributed across multiple sites like Fort Bliss, White Sands Missile Range (WSMR), Fort Irwin, and Twenty-Nine Palms; in some cases, exercises involved road movements between installations. It is unclear whether these latter exercises regularly provided the same types of training opportunities for supporting brigade headquarters that appear evident in the REFORGER and TEAM SPIRIT series. At a minimum, they provide interesting examples of the joint force using multiple, large training complexes across the desert southwest to support large-scale FTXs.

Between the late 1980s and mid-1990s, these large-scale FTXs were either discontinued outright or downsized in favor of smaller exercises focused on CPXs supported by computer simulations (similar to the BCTP). By the latter 1990s, peacetime field training for supporting brigade headquarters was limited to what could be accomplished at home station or much smaller-scale joint and combined exercises.

Training Opportunities Prior to the Start of ODS and OIF

Both ODS and OIF occurred with warning prior to the start of LSCO. This, in turn, provided at least some deploying formations with time to prepare, whether before or after arrival in theater. In particular, some supporting brigade headquarters were able to participate in live,

---

[3] BCTP, "Ops Grps Sierra and Foxtrot: Training Support/Functional/Theater Bdes," briefing slides, undated.

large-scale training events and exercises that helped prepare them to operate and exercise mission command in LSCO.

In ODS, for example, XVIII Airborne Corps arrived months before the start of the air campaign that commenced in January 1991. Its elements had substantial time to train and prepare in theater. Although specific details are unclear as to how individual divisions and brigades trained, at a minimum, the scale and complexity of training opportunities must surely have surpassed those available to supporting brigade headquarters during peacetime. In contrast, VII Corps was only notified for deployment in November 1990, and it arrived in Saudi Arabia in piecemeal fashion starting in December—with some units arriving more than a week after the air campaign began on January 17, 1991. Nonetheless, many units still had several weeks to conduct precombat training in theater before the start of the ground campaign. In addition, VII Corps used the movement from its tactical assembly areas as a major rehearsal that allowed them to test their processes and systems and to make adjustments before combat operations started.

V Corps was the single Army corps committed to OIF. Before deployment, V Corps conducted a series of major command post and FTXs to support planning and to provide opportunities for rehearsals. Because of the timing of the force flow, the 3rd Infantry Division (ID) was the only Army division that was fully deployed well in advance of the start of LSCOs. Once in theater, the 3rd ID had the opportunity to conduct substantial precombat training, using the stockpiled resources and new range complexes constructed after ODS.

In the future, there might be cases in which the Army could benefit from similar opportunities to engage in substantial precombat training. However, one should not count on this— rapidly unfolding contingencies could occur with little or no warning. The Army's operating concept, *The U.S. Army in Multi-Domain Operations 2028*, emphasizes that forces must be ready to deploy and immediately enter the fight in a fast-moving, contested, antiaccess and area denial environment. This means that the quality and rigor of peacetime training will go a long way toward setting the conditions for success or failure in first battles.

MCTP and the WFX

This section describes training opportunities associated with MCTP, particularly the WFX. In short, we find that MCTP and the WFX provide a critical training opportunity, at least for the types of F/MF brigade headquarters that it is designed and resourced to support. However, there are also certain capability and capacity gaps that, if rectified, would further enhance the value of the opportunities that MCTP and the WFX provide. We offer a variety of recommendations in this regard.

The TRADOC regulation governing MCTP defines the WFX as

> a distributed, simulation driven, multiechelon, tactical command post exercise fought competitively against a live, free-thinking adversary . . . . WFXs are directed by the CSA, scheduled by FORSCOM, and conducted by the MCTP.[4]

Across the various stakeholders we spoke with, the consensus was that—apart from mission rehearsal exercises (MREs) that units participate in before deployment or before assuming

---

4   TRADOC, *Mission Command Training Program*, Fort Eustis, Va., TRADOC Regulation 350-50-3, June 23, 2014.

a certain operational mission—the WFX is the most significant mission command training event available to corps and division headquarters and the various types of F/MF brigade headquarters that MCTP supports (we later describe which types of F/MF brigade headquarters MCTP supports).

MCTP includes eight separate OPSGRPs, each of which focuses on a different training audience or performs another function:

- **OPSGRPs A and D**: division and corps headquarters
- **OPSGRP S**: sustainment headquarters—i.e., SUST BDEs and ESCs
- **OPSGRPs B and F**: F/MF brigade headquarters other than sustainment
- **OPSGRP C**: F/MF brigade headquarters other than sustainment at certain WFXs; otherwise, focuses on separate Brigade Warfighter Exercises (BWFXs) for ARNG BCTs
- **OPSGRP J**: special operation forces headquarters
- **OPSGRP X**: development and provision of mission command for exercises.

MCTP currently has the capacity to execute five WFXs per year. These are in addition to supporting five Army service component command (ASCC) exercises per year and six BWFXs per year.[5] BWFXs focus on ARNG BCTs and are distinct from the multiechelon WFXs focused on corps, division, and F/MF brigade headquarters. Note that, when we refer to WFXs, we are *not* referring to BWFXs.

The capacity of each WFX is as follows:

- two corps and/or division headquarters
- two ESCs and/or SUST BDE headquarters (there have been historical incidences of up to three total)
- four to five F/MF brigade headquarters of the following types: aviation, fires (including DIVARTY), engineer, MP, and MEB headquarters (there have been historical incidences of up to six total).

If a given type of F/MF brigade headquarters is not listed above, *then it is not currently a WFX training audience.*

As an example, Figure 3.1 shows the WFX calendar projected for fiscal year 2018, including the units scheduled to participate as training audiences.[6] The window for six scheduled BWFXs is shown in gray, but the timings and training audiences for BWFXs are not specified.

Each month in which a WFX was scheduled is highlighted in green. The row labeled "Exercise" shows the number of the specific WFX (e.g., WFX 18-1) and the dates on which it was scheduled to occur. Subsequent rows show the training audiences scheduled for each WFX. The OPSGRP that supports a given training audience is shown in the leftmost column (OPSGRPs J and X are not shown). Participating division and corps headquarters are underlined; other training audiences are not underlined. Regular Army units are shown in gray, RC units

---

[5]   OPSGRPs A and S also support ASCC exercises (in months with no WFX). Supported ASCC exercises include VIBRANT RESPONSE, ULCHI FREEDOM GUARDIAN, JUDICIOUS RESPONSE, SABER JUNCTION, AUSTERE CHALLENGE, LUCKY WARRIOR, TALISMAN SABRE, and LION FOCUS.

[6]   Figure 3.1 shows the projected schedule as of September 2017. It is possible that ultimately there were changes to some of the units that participated as training audiences. As the figure is only intended to provide a general illustration of what an annual WFX calendar looks lie, any ultimate changes are not consequential.

Figure 3.1
Fiscal Year 2018 WFX Calendar

	OCT	NOV	DEC JAN	OCT	MAR	NOV	MAY	OCT	JUL AUG SEP
Exercise	WFX 18-1 03–12 OCT	WFX 18-2 08–17 NOV		WFX 18-3 06–15 FEB		WFX 18-4 03–12 APR		WFX 18-5 05–14 OCT	
OG-A	I CORPS	2 ID with 2 ID DIVARTY		101 ABN		2 ID		34 ID	
OG-D	42 ID	10 MTN		4 ID		XVIII CORPS 3 UK DIV		1 CD	
OG-S	143 ESC 108 SUST BDE	19 ESC 2 Id SUST BDE		4 ID SUST BDE 101 SUST BDE		3 ESC 321 SUST BDE		55 SUST BDE 1 CD SUST BDE	
OG-F	158 MEB	2 ID CAB 210 FA BDE		36 EN BDE 101 CAB		1 ID CAB 1 ID DIVARTY		225 EN BDE 1 CD DIVARTY	
OG-B	28 CAB 142 FA BDE	130 EN BDE 210 FA BDE		4 ID CAB 204 MEB		926 EN BDE 333 MP BDE		34 CAB 45 FA BDE	
OG-C	194 EN BDE	10 MTN DIVARTY		110 MEB 89 MP BDE		BWFX Window			

SOURCE: Kimo Gallahue, "Mission Command Training Program Overview Brief," briefing, February 10, 2017.
NOTE: MTN = mountain division; ABN = airborne division; AA = air assault; UK = United Kingdom; CD = cavalry division.

are in blue, and one allied participant is shown in red. For example, WFX 18-4 was scheduled for April 3–12, 2018. Although not shown, units were scheduled to participate either from Fort Bragg, North Carolina, or from Fort Riley, Kansas.[7] OPSGRP A was scheduled to support 1st ID headquarters, while OPSGRP D was scheduled to support both XVIII Airborne Corps headquarters and 3rd United Kingdom Division headquarters. OPSGRP S was scheduled to support 3 ESC (a Regular Army unit) and 321 SUST BDE (an RC unit). OPSGRP F was scheduled to support 1 ID CAB and 1 ID DIVARTY (both Regular Army), and OPSGRP B was scheduled to support 926 EN BDE and 333 MP BDE (both RC). OPSGRP C did was not scheduled to support this WFX, as it occurred during the window when OPSGRP C supports BWFXs. For a unit participating as a training audience in a given WFX, the exercise operates on a one-year planning cycle, including three planning events, a one-week Mission Command Training (MCT) session focused on academics (i.e., to review doctrine, the operations process, exercise sustainment and best practices, etc.), and the culminating ten-day WFX.[8] Figure 3.2 illustrates the WFX event lifecycle.

In terms of frequency of WFX participation, *divisional* F/MF brigade headquarters in the Regular Army attend WFXs somewhat more frequently, on average, compared with *non-divisional* brigade headquarters because the former often but not always attend with their parent division headquarters.[9] In some cases, divisional brigades are unable to attend with their parent

---

[7] All WFXs are distributed events involving more than one installation.

[8] One MCTP interviewee stated that key unit personnel first go to MCTP near the start of the one-year cycle for an initial planning conference to talk through the exercise and related processes. Next, MCTP personnel go to the unit for a midyear planning conference. Unit personnel later return to MCTP for a week of MCT academics, during which they focus on topics related to doctrine as well as the coming exercise. Finally, unit personnel return to MCTP for a final planning conference, during which all of the details of the exercise are verified.

[9] The primary command relationship for units based in the continental United States (CONUS) is *attached* (see Appendix A for an overview of Army command and support relationships). In other words, individual CONUS-based F/MF brigade headquarters are attached to either a division or corps headquarters for purposes of training readiness authority, which

Figure 3.2
WFX Event Lifecycle

SOURCE: Gallahue, 2017.
NOTE: MDMP = Military Decisionmaking Process.

division because of such circumstances as operational deployment. In these cases, a brigade might or might not be able to attend a WFX with a different division or corps headquarters.

Figure 3.3 describes the organizational structure of a WFX.

F/MF brigade headquarters do *not* attend WFXs with a complement of full subordinate units—e.g., full battalions, or at least full battalion headquarters—over whom they exercise mission command. As described above, the WFX is a distributed, simulation-driven CPX. F/MF brigade headquarters that participate in the WFX as training audiences exercise mission command over response cells, with which they interact through the simulation (the response cells for F/MF brigade headquarters are described as "work cells" in Figure 3.3). In other words, division and corps headquarters exercise mission command over a mixture of subordinate units (i.e., the F/MF brigade headquarters participating as a training audience) and response cells (generally representing either BCTs or other F/MF brigade headquarters not participating as training audiences). F/MF brigade headquarters, on the other hand, exercise mission command exclusively over response cells. Training audiences are responsible for organizing and sourcing the personnel required for the response cells and for ensuring that they are prepared to support the exercise. As such, response cells tend to be composed of soldiers attached to units subordinate to the F/MF brigade headquarters—but the response cells are not manned and do not operate as whole units (e.g., as whole battalion headquarters), and they do not exercise mission command over any subordinate entities (e.g., companies, teams, or detachments) as part of the exercise.

Key Findings

Based on our interviews with various stakeholders, we find that F/MF brigade headquarters participating in the WFX as training audiences gain valuable mission command experience. This includes experience with:

---

is the authority of a higher-echelon commander to provide direction to an attached unit for leader development, individual and collective training, and unit readiness. We define *divisional* brigade headquarters as brigade headquarters that are attached to division headquarters for training readiness authority; attached F/MF brigade headquarters are by definition not organic to division and corps headquarters.

Figure 3.3
Organization Structure for a WFX

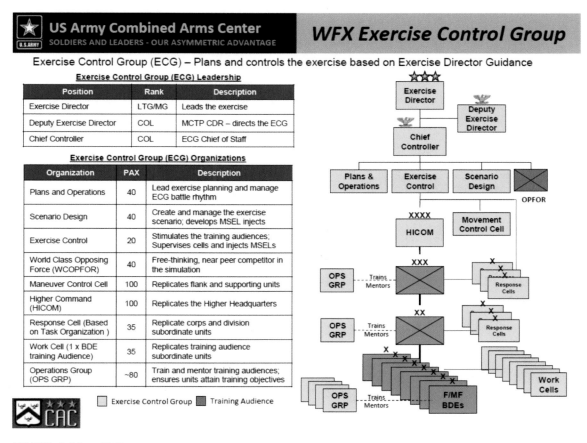

SOURCE: Gallahue, 2017.

- integrating and coordinating with higher, parallel, and subordinate echelons
- establishing and modifying battle rhythms
- testing and refining standard operating procedures (SOPs)
- information management, communication, and use of mission command systems
- establishing and moving the command post.

With regard to integrating and coordinating with higher, parallel, and subordinate echelons, interviews indicated that the WFX is really the only training environment that allows such significant multiechelon integration and coordination; this is generally not possible at home station, and even most joint and combined exercises cannot provide comparable opportunities for brigade headquarters to integrate with parallel and higher echelons.[10] One interview revealed that a significant benefit of the WFX is that it supports standardization across

---

[10] For example, one interviewee found that home station training provided good opportunities to synchronize with subordinate battalions. The benefit of a WFX is that it provides MCT academics, as well as a higher echelon that provides orders to which the brigade headquarters has to respond. Another interviewee expressed the view that, although the WFX was not perfect, it was helpful. Unless a unit deploys, the WFX is the only real way to train mission command.

divisions—e.g., in cases where an F/MF brigade attached to one division headquarters while at home station is attached to different division headquarters to support an operational deployment (although the interviewee said that synchronizing communications in such cases remains a significant challenge).

Nonetheless, interviews highlighted some constraints, some of which might limit the effectiveness of the WFX in certain ways. We have grouped those issues as either capability or capacity constraints. *Capability constraints* relate more to limitations inherent in the WFX as a simulation-driven, tactical CPX focused on headquarters at brigade level and above. *Capacity constraints* relate more to the current limitations of the WFX's resources; these limitations can limit its frequency, duration, and the level of interaction with training audiences. An identified limitation might have both a capability and a capacity dimension. Although we define each identified limitation as primarily a capability or a capacity issue, we discuss both dimensions where relevant. The following section focuses primarily on describing the constraints we identified, and recommendations follow later in this chapter.

As we noted in the introduction, our discussion with MCTP occurred relatively early in the project, before we had the opportunity to conduct unit interviews. Many of the issues we highlight below were based on insights gleaned from unit interviews. Unfortunately, we were unable to meet with MCTP personnel for a second time at the end of the project to get their perspectives or to elicit additional information.

WFX Capacity Constraints

We identified the following capacity constraints, which we discuss in more detail:

- MCTP currently has the capacity for five WFXs per year.
- Some F/MF brigade types, such as TTSBs and E-MIBs, are not eligible to be WFX training audiences.
- MCTP staff are limited in the extent of their interaction with training audiences prior to a WFX.
- WFXs are only ten days in duration, which limits time available to focus on brigade headquarters' training objectives.
- Only some WFXs have joint or multinational content; those that do tend to concentrate this content at corps- and division-level.

MCTP currently has the capacity for five WFXs per year. Again, these are in addition to supporting five ASCC exercises and six BWFXs per year. According to MCTP staff, little if any capacity for more WFXs exists within current resources. One MCTP interviewee stated that, if MCTP were to attempt more WFXs, it would require cutting back on planning time (i.e., planning time and MCT academics). However, this planning is necessary to prepare training audiences for the WFX. We note MCTP's current inability to hold more than five WFXs per year as a constraint, but it is not a constraint that inherently limits the effectiveness of MCTP or the individual WFXs it conducts. It does mean, however, that increasing the number of WFXs per year to train more or different types of units appears infeasible without significant additional resources.

Some F/MF brigade types, such as TTSBs and E-MIBs, are not eligible to be WFX training audiences. MCTP lacks the capacity to include all types of F/MF brigade headquarters as training audiences. Current F/MF brigade training audiences include sustainment, aviation, fires (including DIVARTY), engineer, MP, and MEB headquarters. E-MIBs and TTSBs are examples of functional brigade types that are not included as a training audience (although these brigade types might experience limited participation by resourcing response cells or by augmenting the capabilities of division and corps headquarters when the latter participate as training audiences). Without a significant expansion of resources, it appears impossible to include additional types of functional brigade headquarters as WFX training audiences. Moreover, there is likely also a capability constraint as well as a capacity constraint. We assume that including new functional brigade types would require additions (e.g., OC/Ts with different skills, changes to the scenario and/or simulation to address the training objectives of the new brigade types) to the capabilities that MCTP currently possesses.

MCTP staff are limited in the extent of their interaction with training audiences prior to a WFX. As described, pre-WFX interaction is limited to three planning conferences and a week of MCT academics. MCTP personnel are not involved, for example, in pre-WFX CPXs or other training events that units conduct in preparation for a WFX. One MCT interviewee noted that a unit's home station preparation is crucial to success at the WFX; however, operational tempo for MCTP OC/Ts constrains their ability to interact more with the training audiences. This interviewee found that units would benefit from more interaction, but that it simply was not feasible given time constraints. Although time might be the major constraint, increased MCTP manning might, for example, provide additional resources to support more pre-WFX engagements with training audience units.

WFXs are only ten days in duration, which limits time available to focus on brigade headquarters' training objectives. Some interviewees found that a WFX provided insufficient time to focus both on the training objectives of division and corps headquarters, as well as on the training objectives for participating F/MF brigade headquarters. As such, the former received priority over the latter. The consensus was that MCTP tried to meet brigade training objectives to the degree feasible, but when there were time, resource, or other conflicts, then division and corps headquarters received priority. One interviewee offered the key insight that the WFX simply is not conducive to allowing F/MF brigade headquarters to train to failure, as this would put the training objectives of division and corps headquarters at risk. Some units offered that it might prove beneficial to extend the duration of the WFX to allow a period of increased emphasis on brigade training objectives. Unfortunately, our interview with MCTP did not include the question of whether they had the capacity within current resources to stage a WFX that is longer than ten days, even if just periodically. After this topic was raised during unit interviews, we were unable to reengage MCTP staff to get their perspectives.

Only some WFXs have joint or multinational content; those that do focus it more at corps and division level. Team-building and coordination with UAPs can be a particularly challenging aspect to train. Units often have few if any opportunities at home station. Joint and multinational exercises provide an opportunity for the limited number of units able to participate. On the other hand, in the context of an LSCO, there is not uniform requirement across F/MF brigade types to team-build and coordinate with UAPs. The need to do so will vary based on the specific scenario (which in part will determine the type and characteristics of UAPs that are relevant) and the specific type of F/MF brigade headquarters (different brigade headquarters within the same type might have different requirements based on specific missions). Regardless, WFXs, as currently constituted, provided F/MF brigade headquarters with little (if any) opportunity to exercise with UAPs. This is not inherently a gap. However, if the intent of the WFX was to help prepare specific units to execute specific missions in the context of specific LSCO scenarios (a topic we return to later), then this limitation has increased relevance.

WFX Capability Constraints

We identified the following capability constraints for F/MF brigade headquarters, which relate to limitations inherent in the WFX as a simulation-driven, tactical CPX focused on headquarters at brigade level and above.

- Training value from interacting with higher and parallel echelons at the WFX can be greater than value from interacting with subordinate echelons.
  - For units identified for sourcing for an OPLAN, there would be value in a WFX reflecting the OPLAN task organization.
- As a simulation-driven CPX, the WFX—although highly beneficial—cannot fully match the training value of operating in a field environment.
  - For some F/MF brigade types, WFX simulation is not designed to stress all their METs.
  - The nature of the simulation might provide unrealistic expectations of how processes will work in actual operations.
  - Interacting with battalion response cells can provide less training value than interacting with subordinate battalions operating in the field.

Training value from interacting with higher and parallel echelons at the WFX can be greater than value from interacting with subordinate echelons. For F/MF brigade headquarters, interviewees suggested that the training benefit from interacting with higher and parallel echelons was often greater than the benefit from interacting with subordinates[11]—although exercising command over subordinates is a key element of mission command training. Interviews noted that brigade headquarters can become focused on coordinating with higher echelons, at expense of coordinating with subordinates.[12] In the simulation, subordinates are simply response cells waiting to interact with the brigade headquarters, rather than actual subordinate units executing tasks in field conditions. This problem is exacerbated by the fact that brigade headquarters are sometimes unable to adequately prepare their response cells to execute their exercise function.[13] (Some suggested that greater assistance from MCTP

---

[11] At home station, most SBs are attached to a division headquarters. In the WFX, however, SBs are aligned under ESCs rather than under a division, to reflect normal doctrinal command relationships in an operational environment. Some WFXs do not include an ESC as training audience, but include only a HICOM cell representing the ESC, which might not even be sourced with personnel from an ESC (i.e., the ESC HICOM cell might be sourced using personnel from a different SB). One SB we interviewed indicated that, if the WFX includes only an ESC HICOM cell, then the cell might be unable to run a full battle rhythm or might produce mission orders that are not sufficiently detailed to exercise the SBs response—adversely impacting the SBs training value.

[12] For example, one interviewee noted that communicating and coordinating with response cells could be a challenge. Brigade headquarters could get caught up in coordinating with higher echelons and forget to also filter information down.

[13] For example, one interviewee noted that the preparation of the unit elements forming the response cells, and the quality of their preparation, has a big impact on training value. This means that it is an important task to ensure that subordinate units providing personnel for the response/work cells are ready. Otherwise, there is a steep learning curve at the WFX for the response/work cells that can impact the quality of the training experience for the training audiences. Another noted that it takes between some and a significant amount of training for response cells supporting the simulation to be prepared to provide quality feedback.

in this regard could prove highly beneficial, but as discussed earlier, MCTP currently lacks the capacity to do so.)

On a related point, regarding units identified for sourcing on a specified combatant commander OPLAN, we asked units if there would be benefit in having a WFX that reflected the OPLAN's task organization. In other words, units could be selected for participation at such a WFX—including corps, division, ESC, and F/MF brigade headquarters—to reflect at least a portion of the task organization indicated in the OPLAN's time-phased force deployment data (TPFDD), which include information indicating the combatant commander's intended task organization once units have arrived in theater. The consensus view among the units we interviewed was that this approach should be quite valuable. However, there would also be a challenge if the OPLAN task organization did not match home station alignments, particularly for divisional brigade headquarters. For example, there would be a challenge if a CAB was expected to serve under a different division as part of the OPLAN, rather than the "parent" division to which it was normally attached when at home. It would likely prove too taxing to both support the parent division for its WFX and to attend a TPFDD-focused WFX within the same two-year cycle. Moreover, divisional brigade headquarters in particular indicated that support from the parent division—both in terms of dedicating resources and fencing off time for training—was important for successful WFX preparation; such support was generally easier to get when the brigade headquarters attended the same WFX as the parent division.

As a simulation-driven CPX, the WFX—although highly beneficial—cannot fully match the training value of operating in a field environment. Although the CABs, SUST BDEs, and EN BDEs we spoke with found value in WFX participation, there was a general consensus that a true gold standard training opportunity for LSCO required training in a field environment. For example, although not the case for every single unit we spoke with, we heard from at least some CAB, SUST BDE, and EN BDE interviewees that the WFX simulation was not capable of stressing all their METs. This did not mean that interviewees did not find value in the exercise, only that some METs could only be fully exercised in a field environment. Similarly, TTSB and E-MIB interviewees generally expressed the view that, even if they were able to participate in the WFX as training audiences, the nature of the simulation was not suitable to exercise their METs.

Some interviewees felt that the nature of simulated environments had the unintended consequence of providing unrealistic expectations of how processes will work in actual operations. For example, one SUST BDE noted that changes in task organization at the WFX are too simplified; they happen faster at the WFX than in a field environment and do not account for the needs (e.g., maps, overlays) and limitations of subordinate echelons.

There was a general consensus among units that we interviewed that interacting with battalion response cells at the WFX provides less training value than interacting with subordinate battalions operating in the field. Some noted that, although the WFX provided the best opportunity for integrating with parallel and higher headquarters in a training environment, certain FTXs—particularly major Army, joint, or combined exercises, as opposed to home station activities—provided better training opportunities for exercising mission command over subordinate echelons.[14]

---

[14] For example, DYNAMIC FRONT is a combined exercise that promotes fires interoperability down to the battery level among North Atlantic Treaty Organization (NATO) partners. One interviewee expressed the view that there was no better way to train for mission command over subordinates than to attend exercises like DYNAMIC FRONT. This interviewee

When asked what would constitute a true gold-standard training opportunity for LSCO, several of the CABs and EN BDEs we spoke with stated that they would like to receive a dirt CTC–like experience. For example, one EN BDE interviewee stated that—other than general engineering operations involving tasks such as construction planning—no live events right now really exercise an EN BDE in areas like mission command for a major breaching or wet gap crossing. Such exercises would require a dedicated CTC-like event, with the EN BDE as the primary training audience. Another EN BDE interviewee noted that, to fully test an EN BDE's proficiency to command a wet gap crossing, an exercise would need multiple crossings and division- or corps-level integration to really stress training with maneuver elements. Although there are smaller-scale joint and multinational exercises that incorporate wet gap crossings that are beneficial to a degree, they simply do not occur at the same scale and so do not fully stress the EN BDE's capability.

Similarly, most of the CABs we spoke with indicated that they would like to receive a CTC-like event as the supported training audience. One stated that, ideally, the CAB would also have the opportunity to integrate and exercise in a field environment with a division headquarters, a FA BDE and/or DIVARTY, and relevant elements of the Joint force (e.g., to exercise Joint air attack team [JAAT] capabilities).

No SUST BDEs we interviewed expressed the desire for a CTC-like event per se. However, one SUST BDE interviewee found that a gold-standard training event would provide the opportunity to practice distribution operations over distance, with convoy operations multiple times per day; doing so would require spaced-out units to support. An interviewee from the Combined Arms Support Command (CASCOM) stated that the SUST BDE should ideally receive a major training event focused both on mission command and on the SUST BDE's broader METs—one that should take place in a wartime context (e.g., with HICOM exercised by an ESC). According to the interviewee, the event should "throw the kitchen sink to really stress the staff"; however, doing so is not conducive to a WFX focused on training needs of the division and corps headquarters.

Interviewees offered some other suggestions for an enhanced WFX focused more on the training needs and objectives of F/MF brigade headquarters:

- **Units should be forced to actually "deploy" to an installation.** For example, one CAB interviewee identified the need to practice more with speed of deployment and assembly ("in practice, we move lethargically"). Other CABs made similar observations on the value of practicing deployment operations, as did some SUST BDEs, although they generally noted that they had other opportunities in this regard.
- **Make command post jumps more often, up to every 24 hours.** One SUST BDE interviewee argued that, in actual LSCO, the digital signatures of command posts make them "sitting ducks"; at a minimum, brigade headquarters need to practice moving much more often. The interviewee also noted that reductions in headquarters personnel authorizations over the past decade make jumps more difficult.[15]

---

found the WFX to be better for practicing integration and communications with the division echelon but inferior with regard to exercising mission command over subordinates.

[15] Similarly, an interviewee from the Fires CoE noted that brigade headquarters might also have challenges in executing 24-hour operations, given reductions to authorized manning levels. This interviewee stated that, during the OIF/OEF era,

- **Exercise mission command in a communications-degraded environment.** For example, an interviewee from the Maneuver CoE stated that the Army has become "drunk on information"; the Army assumes it will have this information in large-scale operations, but Army systems can be compromised by an adversary. The Army must practice mission command without all this information.

In summation, the WFX is a highly valuable training event for those F/MF brigade headquarters that participate. In general, short of an MRE prior to a known deployment or similar mission, the WFX provides the Army's premier training opportunity for F/MF brigade headquarters that participate. However, the WFX has capability and capacity constraints that place limits on its effectiveness in meeting the training objectives of participating F/MF brigade headquarters. For those brigade headquarters that participate, their training objectives are secondary to those of participating division and corps headquarters. In particular, *the WFX generally does not permit F/MF brigade headquarters to train to failure*, as this would risk the ability of divisions and corps to meet their training objectives. Moreover, some METs for some brigade types cannot be effectively trained at a WFX because of the constraints inherent to the simulation-driven CPX environment. In short, despite the significant value that the WFX provides, limitations mean that training might fall short of a true gold standard for LSCO.

Potential Risks If Brigade Headquarters Do Not Receive Gold-Standard Training Opportunities

What are the consequences if F/MF brigade headquarters do not receive true gold-standard training opportunities? What are the potential risks to force and mission in that case? These are important questions when deciding to invest significant resources to further enhance training opportunities.

Unfortunately, we found it difficult to assess the potential risk to force and mission resulting from these limitations. Brigade headquarters that we interviewed generally estimated that risks were not significant and that they could be overcome in a relatively short period of time at the start of an operation.

For example, units we interviewed reported that, in an operation, they would face challenges integrating with and/or exercising mission command over units in theater with which they had not habitually trained. The challenges they commonly reported related to synchronizing SOPs and communications, as well as time needed to build personal relationships. An interviewee from the Fires CoE provided perhaps the most thorough statement, articulating the particular challenges related to integrating fires as follow (comments are paraphrased rather than a direct quote):

> FA BDEs and DIVARTY must integrate and deconflict with CABs, air defense, and Air Force units. Not all these capabilities will be nested or integrated on a routine basis, so there is the potential for fratricide if the division headquarters has not clearly articulated a plan to subordinate echelons. Synchronization with staff and across echelons is the challenge. It is resource intensive to do this outside of a WFX. There are many participants in the chain. It is a very complex process. One element that is off throws the entire system off. The Army had a lot more sets and reps before OIF. Systems have become more complex. In LSCO (as

---

there was a focus on finding efficiencies in headquarters to build more units of types in higher demand; for LSCO, headquarters needed more personnel overhead for 24-hour operations.

opposed to OIF), brigades are in mobile tactical operations centers without assured power and are continually jumping. There is risk. The Army is attempting to address this through WFXs and multi-national exercises. However, without a division construct—i.e., training as a warfighting division that controls its warfighting package—everything is going to be a pick-up game. Every division is different. Modularity hinders echelon above brigade mission command. If you really want a good comparative analysis, execute a WFX with no notice. That will give the answer of whether the system is working.

Among units we interviewed, the consensus was that the challenges related to integrating across echelons in the absence of established relationships would likely lead to some initial decrease in effectiveness, as well as some risk to force and mission. However, there was no consensus that risks were significant, and most interviewees believed that challenges should be overcome in, at most, a few weeks.

Similarly, although many units we spoke with noted that they experienced challenges and limitations in training all their wartime METs, the consensus was that the challenges could be overcome in the short term. There was no consensus that risks to force or to mission in an LSCO were significant. One interviewee from an EN BDE made perhaps the strongest characterization of risk, stating that the inability to train to command a large-scale wet gap crossing on a combined arms basis with higher echelons present constituted risk both to force and to mission in a way that is a "fall-on-sword issue." However, an interviewee from another EN BDE did not believe such risks were as significant, finding that the experience gained as a planner who had worked in many different jobs over a career had provided the interviewee with a systematic understanding of how different capabilities operate and interact; this would enable the ability to surmount challenges with limited risk.

We note, however, that although the F/MF brigade headquarters that we interviewed generally estimated that overall risks were not significant, most personnel have focused on COIN operations since 2003; they might be limited in their ability to fully assess risk for LSCO.

Recommendations

On the one hand, limitations associated with current training approaches indicate that training falls short of a true gold standard for LSCO. On the other hand, the consensus among unit staff that we interviewed was that risks to force and mission were not significant and could be overcome in the near term. As such, we do not recommend wholesale changes to current training approaches applicable to all F/MF brigade headquarters. Instead, we offer recommendations that focus on the themes of prioritization and experimentation. In short, we recommend that the Army *experiment* with an enhanced WFX, possibly with a major field training component, to assess whether the value that these enhanced training opportunities provide justify any increased costs. For these enhanced WFXs, we recommend that the Army *prioritize* specific F/MF brigade headquarters most in need of enhanced training opportunities. If the results of these experiments indicate that they provide significant training value that justifies increased costs, the Army could consider implementing these recommendations on a more systematic basis.

As noted in the introduction, we were unable to reengage with MCTP to get its perspective on our recommendations. If the Army chooses to move ahead with recommendations that relate to MCTP or the WFX, it should ensure that MCTP has an opportunity to review and provide input before implementation. These recommendations were largely developed using input from soldiers assigned to F/MF brigade headquarters who had either participated in, or were preparing to participate in, a WFX. Nonetheless, input from MCTP and other personnel experienced with the design and execution of simulated and live exercises will be necessary to fully determine the feasibility of these recommendations—whether in terms of the resources required to implement them or in terms of related challenges, constraints, or concerns that could adversely affect the training value that the recommendations are intended to provide.

Recommendation Three

Focus enhanced training opportunities on specific F/MF brigade headquarters with highest priority LSCO missions. Although limitations associated with current training approaches might mean that training falls short of a true gold standard for LSCO, not all F/MF brigade headquarters necessarily need to receive gold-standard training opportunities for every cycle. Enhanced training opportunities are most important for units identified for OPLAN sourcing and aligned against key OPLAN missions. In particular, the Army should prioritize OPLANs that might occur with little or no warning and would provide little or no time in theater for precombat training. If the OPLAN includes more units than the Army can provide enhanced opportunities for, then the Army should focus on those units with the most important OPLAN missions. As part of assessing which units have the most important OPLAN missions, the Army should consider whether any other types of brigade headquarters warrant inclusion, in addition to the types normally included as a WFX training audience.

Having identified priority units, the Army should schedule certain WFXs to match key portions of the OPLAN task organization. Many of the brigade headquarters we interviewed stated that OPLAN-focused WFXs would be beneficial.

In short, even if the Army adopted only this recommendation, our interviews indicate that holding OPLAN-focused WFXs would provide enhanced value—particularly if certain elements of TPFDD task organization matched the home station attachments of F/MF brigade headquarters to division and corps headquarters, as this would better enable synchronization and support of pre-WFX preparations across echelons.

The following two recommendations offer additional ways to enhance the training value of specific, OPLAN-focused WFXs. The first recommends experimenting with longer-duration WFXs to enable increased focus on brigade headquarters' training objectives, potentially with more pre-WFX interaction/mentoring between units and MCTP staff. The second recommends experimenting with adding a field training component to WFXs that includes subordinate formations rather than simply response cells. In other words, Recommendations Three, Four, and Five are intended as a series of recommendations in support of a common goal. However, the choice to only implement Recommendation Three—or only Recommendations Three and Four—should still provide enhanced value over current approaches.

Recommendation Four

For prioritized units, consider experimenting with a longer-duration enhanced WFX, adapted to better meet the training goals of F/MF brigade headquarters. An enhanced WFX could be extended in duration—e.g., from the current ten days to 15 or 20 days—to provide more time to focus on training objectives of brigade headquarters.[16] For example, the first ten days could focus on the training objectives of division and corps headquarters;

---

[16] The current ten-day WFX is made up of two five-day cycles, each ending with an after-action review (AAR).

the remaining five or ten days could focus on the training objectives of F/MF brigade headquarters. However, better ways to structure the exercise might exist. We do not offer specific recommendations as to how long an extended WFX should be, or how training focused on different audiences should be sequenced with the WFX; Army training planners should make those decisions. The overall point is that the Army should consider experimenting with a longer-duration WFX so that at least a portion of the exercise can focus on the training objectives of F/MF brigade headquarters, providing them an opportunity to train to failure without adversely affecting the training objectives of division and corps headquarters. As part of this recommendation, the Army should assess the feasibility of increased pre-WFX interaction and mentoring between training audiences and MCTP staff.

Implementing this recommendation as an experiment—without a significant increase in resources available to MCTP—might require that two existing WFX events be merged into a single event. In other words, for the purpose of an initial experiment, it might be necessary to cancel one of the five annual WFXs, so MCTP can focus resources on one of the remaining four to stage it as an enhanced WFX. The Army should also assess whether there are capabilities already resident within the Army that could be surged to increase MCTP capacity for an enhanced WFX. For example, USAR units that support First Army might be able to provide capabilities that can be surged to support an enhanced WFX.

Implementing this recommendation need not inherently change the current WFX design as a simulation-driven CPX. Keeping this format but extending it in duration to allow additional time to focus on brigade training objectives should add value. However, as discussed above, the current format presents some limitations in meeting brigade headquarters training objectives

Recommendation Five

Experiment with including a field training component as part of an enhanced WFX. Most of the CAB, SUST BDE, and EN BDEs staff that we interviewed reported that it would be valuable to receive a dirt CTC–like experience in which they could train to exercise their wartimes METs in a field environment as the supported training audience. Ideally, such an opportunity also would allow the brigade headquarters to integrate with parallel and higher echelons, replicating their expected operations in an LSCO.

One approach to at least partially achieve this objective could be to provide a field training component as part of an enhanced WFX. This would enable brigade headquarters to command battalions operating in field conditions, rather than simply commanding response cells as part of the WFX simulation. One option would be to include battalions and lower-echelon units that are also identified for TPFDD sourcing, thus replicating the TPFDD task organization and providing the battalions and lower echelon a key training opportunity while supporting integration and team-building with the brigade headquarters that are expected to command them when executing the OPLAN.

Such an exercise would likely need to occur on a distributed basis spread across multiple locations and linked by mission command networks and systems. Installations across the U.S. desert southwest are candidates to provide space for coordinated fires, movement, and maneuver. Other locations could be included to meet all training objectives, such as an exercise with a large-scale wet gap crossing. If feasible, this event could also be coordinated with dirt CTC rotations for BCTs identified for TPFDD sourcing within the same portion of the task organization. Although not all BCTs identified for TPFDD sourcing could execute simultaneous dirt CTC rotations, those that could not would still be eligible to participate as response cells responding to division and corps headquarters as part of the exercise. Enhanced joint and allied participation could also be considered.

We assume that the types of F/MF brigade headquarters that currently participate in the WFX would remain the explicit training audience for an enhanced WFX. In other words, the MCTP capacity constraints that we discussed earlier would likely still prohibit the inclusion of such brigade types as E-MIBs and TTSBs as a formal training audience. Nonetheless, at least some of these other brigade types might derive substantial training value from supporting such a large-scale FTX, even if they are not the formal training audience. For example, the TTSB staff that we interviewed noted that one of its greatest challenges involved staging exercises in which they actually had customers to support—ideally, one in which brigade headquarters, battalions, and companies practiced real-world operations and provided the TTSB an opportunity to manage a network that featured multiple ESBs.

Obviously, scheduling an event of this type would be an enormous undertaking. It was beyond the scope of our effort to attempt detailed proposals regarding how such an exercise might be structured and implemented or how much such an exercise might cost. We note, however, that there are historical precedents prior to OIF for such large-scale, multiechelon exercises, as we summarized briefly at the start of this chapter and discuss in more detail in Appendix B.

Recommendation Six

Prioritize unit participation in certain key joint and multinational exercises based on planned TPFDD alignments (if not already doing so). Unit interviews suggested that existing joint and multinational exercises can provide substantial training value. Although they generally do not provide the same types of opportunities to integrate with parallel or higher echelons that the WFX provides, they can provide valuable opportunities to exercise mission command over subordinate echelons, often in field conditions. Currently, they also provide key opportunities for training or otherwise gaining experience with joint and multinational partners.

For participation in such exercises, the Army should consider prioritizing units identified for OPLAN sourcing. In particular, this should apply to the types of F/MF brigade headquarters that are not included as WFX training audiences.

External Evaluations

The final section of this chapter discusses the role of EXEVALs in assessing training proficiency. As described in FM 7-0, *Train to Win in a Complex World*:

> EXEVALs are unit proficiency evaluations. They are formal in nature and conducted external to the unit. The EXEVAL provides commanders with an objective way to evaluate their unit METs or selected collective task proficiencies. *All units in the Army undergo an EXEVAL to validate fully trained (T) or trained (T-) task proficiency ratings.* (italics added for emphasis).[17]

---

[17]   FM 7-0, 2016, p 3-13.

Per FM 7-0, EXEVALs include the following key requirements:

- The higher commander two levels up approves and resources it.
- The commander resources it to achieve a minimum of T or T- task proficiency rating.
- The higher commander (one or two levels up) trains and certifies external OC/Ts. The senior OC/T can be from an adjacent unit within the HICOM of the unit evaluated.
- The higher commander trains and evaluates METs and battle tasks (to include battle drills).
- T&EOs are the objective basis of the evaluation.
- The higher commander two levels up supervises the final AAR.
- The formal commander (one level up) discusses with the unit commander the expected proficiency levels for METs and battle tasks (including battle drills) and overall level of proficiency for readiness reporting units.[18]

In short, EXEVALs are intended to be used to evaluate a unit's METL proficiency. Recall from Chapter Three that no F/MF brigade headquarters has a MET focused exclusively on mission command; instead, "Conduct Mission Command Operations Process for Brigades" is a supporting collective in every brigade MET. In other words, the purpose of the EXEVAL is not directly to evaluate mission command proficiency. However, the ability to exercise mission command in the context of each MET is an element of the unit's proficiency in that MET.

As part of the WFX, MCTP provides units with two formal AARs that discuss unit performance with regard to mission command and other issues that bear on unit METL proficiency. However, MCTP is explicit that these AARs are not formal EXEVALs, that they are not based on unit T&EOs, and that MCTP does not evaluate unit training proficiency for the purpose of determining unit T-ratings.[19]

Key Findings

As part of our interviews with F/MF brigade headquarters, we discussed how EXEVALs factor into their training. *None of the F/MF brigade headquarters that we interviewed reported having received an EXEVAL of their METL proficiency for LSCO.* (EXEVALs should not be confused with predeployment validations of unit proficiency prior to beginning an assigned mission—such as deploying for an ongoing named operation or assuming a homeland response force mission—which some units did report receiving.[20]) In other words, brigade headquarters are rating their own METL proficiency, rather than having a HICOM independently evaluate their training proficiency.

Most of the brigade headquarters staff that we spoke with stated that they would like to receive a formal EXEVAL. However, they were unable to assess the degree to which the current lack of an EXEVAL negatively affects their training. In other words, although most thought that an EXEVAL would be beneficial, none stated that the lack of an EXEVAL had an obvious and significant impact. Moreover, they were generally skeptical that higher headquarters would be able to dedicate the time and resources to provide EXEVALs. In addition,

---

[18]   FM 7-0, 2016, p 3-13.

[19]   Gallahue, 2017. This insight was reinforced during interviews with MCTP personnel.

[20]   We did not specifically ask about validation for non-LSCO missions. In some cases, however, the topic arose over the course of the interview.

one interviewee suggested that some evaluators might no longer know what "right" looks like, indicating a concern that the Army's recent focus on COIN has made both units and external evaluators unable to properly evaluate LSCO skills. Finally, the TTSB staff we interviewed expressed the view that higher echelons lacked the specialized expertise for a proper EXEVAL and that TTSBs should evaluate each other—a process that would aid standardization. This view appears to be compatible with FM 7-0 guidance, which states that "senior OC/T can be from an adjacent unit within the higher command of the unit evaluated";[21] however, in many cases an "adjacent unit" would need to be external to the HICOM, as in most cases there is only a single F/MF unit of a given type per division (in some cases, there is only one per corps).

In addition, some expressed skepticism that any of their current training events provided a proper venue for an EXEVAL. As discussed above, the consensus view was that—apart from MREs that provide validation for specific missions—the WFX is generally the most significant major training event for the F/MF brigade headquarters that participate. However, WFXs as currently constituted do not prioritize the training objectives of F/MF brigade headquarters. Moreover, some unit staff stated that the simulated environment of the WFX was insufficient to train all of their METs. As such, the current WFX does not appear to provide a suitable venue for fully evaluating a unit across its wartime METs. Finally, the WFX as currently conducted is primarily viewed as a learning opportunity rather than a venue to formally evaluate training proficiency. In other words, the AAR process for the WFX is focused on unit self-evaluation and honest feedback from OC/Ts and mentors. If the WFX is transformed into a venue for EXEVAL, its value as a learning opportunity could atrophy.

Recommendation Seven
The Army should provide F/MF brigade headquarters with EXEVALs as provided for in Army training doctrine; the Army should consider whether it is feasible and desirable to associate EXEVALs with enhanced WFXs (for brigades that participate in such exercises); and the Army should consider the advisability of partially resourcing such EXEVALs using a permanent generating force organization to provide OC/Ts. Army doctrine states that all units in the Army must undergo an EXEVAL to validate training proficiency. Most of the brigade headquarters staff that we spoke with stated that they would like to receive a formal EXEVAL, but none reported having received one. We recommend that the Army provide F/MF brigade headquarters with EXEVALs as provided for in Army training doctrine, or change its doctrine if it determines EXEVALs are not warranted or feasible at brigade level.

The Army should also consider whether it is feasible and desirable to associate EXEVALs with enhanced WFXs (for brigades that participate in such exercises). This recommendation is premised in part on the notion that enhanced WFXs can and should be constructed to provide a venue appropriate for an EXEVAL. As noted above, however, this could diminish the value of an enhanced WFX as a learning opportunity. On the other hand, if the Army determines that the WFX is not a suitable venue for an EXEVAL, it is unclear what an alternative venue would be.

The Army would also need to determine which high headquarters should conduct the EXEVAL at an enhanced WFX. For example, should the EXEVAL for a WFX that is TPFDD-focused be conducted by a unit's higher headquarters in the TPFDD task organization (if this

---

[21]   FM 7-0, 2016, p 3-13.

differs from the higher headquarters that exercises peacetime ADCON over the unit)? We assume the expected wartime organization is the preferred option, but the Army would need to make this determination. Moreover, in the case of the SUST BDE, for example, it would appear more suitable to have an ESC conduct the EXEVAL, even if the division to which the SUST BDE is nominally attached in peacetime happens to also be included in the TPFDD. By doctrine, in an operation, an SUST BDE is normally attached to an ESC or theater sustainment command (TSC) rather than to a division.[22]

Finally, the Army should consider whether it is feasible and valuable to resource a permanent generating force organization to provide OC/Ts for brigade-level EXEVALs. This could promote standardization—ensuring brigade headquarters are evaluated based on a common set of objective criteria—and could better ensure that qualified OC/Ts are available to support EXEVALs.

---

[22] Per ADRP 4-0, *Sustainment*: "When deployed, the sustainment brigade is a subordinate command of the TSC, or by extension the ESC. . . . It plans, prepares, executes, and assesses sustainment operations within an area of operations. It provides mission command of sustainment operations and distribution management. . . . (The TSC or ESC) employs sustainment brigades to execute theater opening (TO), theater sustainment, and theater distribution operations. . . . The sustainment brigade normally remains attached to the TSC or ESC but supports the division. The division may have [operational control (OPCON)] of a sustainment brigade while conducting large-scale exploitation and pursuit operations" (ADRP 4-0, *Sustainment*, Washington, D.C.: Headquarters, Department of the Army, July 13, 2012).

CHAPTER FOUR

Summary of Findings and Recommendations

This concluding chapter summarizes our key findings and recommendations.

Findings

The Army provides valuable mission command training opportunities for F/MF brigade headquarters, but limitations might mean training falls short of a true gold standard for LSCO. In particular, MCTP's WFX is a highly valuable training event for the F/MF brigade headquarters that participate. In general, short of an MRE prior to a known deployment or similar mission, the WFX provides the Army's premier training opportunity for F/MF brigade headquarters that participate. However, the WFX has capability and capacity constraints that limit its effectiveness in meeting the training objectives of participating F/MF brigade headquarters.

For example, MCTP lacks the capacity to include all types of F/MF brigade headquarters as training audiences. Current F/MF brigade training audiences include sustainment, aviation, fires (including DIVARTY), engineer, MP, and MEB headquarters. E-MIBs and TTSBs are examples of functional brigade types that are not included as a training audience (although these brigade types might experience limited participation by resourcing response cells or by augmenting the capabilities of division and corps headquarters when the latter participate as training audiences). Without a significant expansion of resources available to MCTP, it appears impossible to include additional types of functional brigade headquarters as WFX training audiences.[1] Given resource constraints, we do not specifically recommend expanding the capacity of MCTP to *routinely* include additional brigade types (see Recommendation Three in the next section for a related issue).

More significantly, for those types of F/MF brigade headquarters that are WFX training audiences, training objectives are secondary to those of participating division and corps headquarters. In particular, *the WFX generally does not permit F/MF brigade headquarters to train to failure*, as this would risk the ability of divisions and corps to meet their training objectives. Moreover, some METs for some brigade types cannot be effectively trained at a WFX (or at home station); some form of field training or other specialized training is required. Large-scale

---

[1] We assume this is more than just an issue of total capacity—i.e., the total number of F/MF brigade headquarters that MCTP can train across its five WFXs in a given year. We assume that including new functional brigade types would also require additions to the capabilities that MCTP currently possesses (e.g., OC/Ts with different skills, changes to the scenario and/or simulation to address the training objectives of the new brigade types). Unfortunately, our discussion with MCTP staff early in the project did not address this with great specificity, and we were unable to meet with them a second time to refine our understanding of this topic.

joint and combined exercises are an alternate training venue, and they can provide significant training value. However, the number and frequency of brigade participation might be limited, and, in general, such events do not provide the same level of integration with parallel and higher headquarters that the WFX provides.

We found it difficult to assess the potential risk to force and mission resulting from these limitations. Brigade headquarters that we interviewed generally estimated that risks were not significant and that they could be overcome in a relatively short period of time at the start of an operation. We note that, historically, such operations as ODS in 1991 and OIF in 2003 occurred with enough warning to provide at least some participating units additional time to train in an integrated fashion (i.e. ,with higher, parallel, and subordinate echelons), sometimes in a field setting. However, such warning time and opportunity to train cannot be guaranteed for all contingencies. Moreover, although the F/MF brigade headquarters that we interviewed generally estimated that risks were not significant, most personnel have focused on COIN operations since 2003, and they might be limited in their ability to fully assess risk for LSCO. No brigade headquarters that we interviewed reported receiving an EXEVAL of their training proficiency for LSCO.

Not all F/MF brigade headquarters need to receive a gold-standard training opportunity for LSCO or to receive such an opportunity every training cycle. Enhanced training opportunities are most important for those units aligned to OPLAN TPFDDs that require rapid deployment on short notice, particularly if there is little expectation that units will have additional time to train together once in theater.

Next, we summarize our recommendations from the preceding chapters. Our first two recommendations deal with home station training and are meant to be generally applicable, at least across brigade headquarters of a given type, without regard to their alignment on an OPLAN. Recommendations Three through Seven focus specifically on brigade headquarters aligned to key OPLANs. *They share a key theme: prioritization and experimentation, as follows:*

- **Prioritize** individual brigade headquarters most in need of enhanced training opportunities.
- **Experiment** with an enhanced WFX, possibly with a major field training component, to assess whether the value that these enhanced training opportunities provide justify any increased costs.

Recommendations

Recommendation One
Across brigade types, encourage and disseminate examples of innovative home station training. Some brigade headquarters we interviewed reported innovative ways of practicing mission command while at home station. In some cases, implementing these methods can require permissions and coordination across multiple stakeholders, including parent division or corps headquarters, a CTC or garrison commander where the event will occur, and the supported training audience (e.g., if conducted in conjunction with a BCT CTC rotation).

We recommend that the Army establish a form to collect and disseminate innovative training examples as part of a lessons learned process. The Army should also consider addressing innovative training examples as part of FORSCOM annual training guidance or a similar forum to help stimulate implementation and coordination among multiple stakeholders.

Recommendation Two

For certain types of brigade headquarters that lack an organic signal company, the Army should study the challenges this causes for home station training, as well as options for mitigation. Some types of brigade headquarters identified the lack of an organic signal company as a capability gap that presents a critical challenge for home station training.[2] Others we spoke with stated that the signal community has the capacity to support home station training for these brigade headquarters, if requested on a timely basis. The issue appears to be that the signal community can support major training—perhaps at higher rates than they are currently tasked—but the brigade headquarters without organic signal companies want the ability to use tactical networks on a regular, unplanned basis. For brigade headquarters that currently lack an organic signal company, we recommend that the Army assess the costs and benefits of providing a sustained signal capability to support home station training, at least for high-priority units identified for TPFDD sourcing. The answer might not be an organic signal company. For example, it might be feasible instead to attach an element organic to an ESB to support the brigade's training on an extended but not permanent basis.

Recommendation Three

Focus enhanced training opportunities on specific F/MF brigade headquarters with the highest-priority LSCO missions. Not all F/MF brigade headquarters necessarily need to receive a gold-standard training opportunity for LSCO, or to receive one every training cycle. The Army should prioritize units identified to source key OPLAN missions and should schedule certain WFXs to match key portions of the OPLAN task organization. As part of assessing which units have the most important OPLAN missions, the Army should consider whether any other types of brigade headquarters warrant inclusion as part of an enhanced WFX, in addition to the types normally included as a training audience.

The following two recommendations offer additional ways to enhance the training value of specific, OPLAN-focused WFXs. In other words, Recommendations Three through Five are intended as series of recommendations in support of a common goal. However, the choice to only implement Recommendation Three—or only Recommendations Three and Four—should still provide enhanced value over current approaches.

Recommendation Four

For prioritized units, experiment with conducting a longer-duration enhanced WFX, adapted to better meet the training goals of F/MF brigade headquarters. The Army could keep the basic WFX format as a simulation-driven CPX, but extend it in duration—e.g., from ten to 20 days—to provide more time to focus on training objectives of brigade headquarters. The Army should also assess the feasibility of increased pre-WFX interaction and mentoring between training audiences and MCTP staff. In addition, the Army should assess whether there are capabilities already resident within the Army that could be surged to increase MCTP capacity for an enhanced WFX.

---

[2] The lack of an organic signal company is not necessarily a capability gap once the brigade headquarters is deployed for an operation, provided the brigade receives the necessary signal support in theater.

Recommendation Five

Experiment with including a field training component as part of an enhanced WFX. The current WFX format as a simulation-driven CPX presents some limitations in meeting brigade headquarters training objectives. The Army should consider adding a field training component as part of an enhanced WFX that would enable brigade headquarters to command battalions operating in field conditions, rather than simply commanding response cells as part of the WFX simulation. One option would be to include battalions and lower echelon units that are also identified for TPFDD sourcing, thus replicating the TPFDD task organization and providing the battalions and lower echelon with a key training opportunity as well.

Such an exercise would likely need to occur on a distributed basis spread across multiple locations and linked by mission command networks and systems. Enhanced joint and allied participation should also be considered. There are historical precedents prior to OIF for such large-scale, multiechelon exercises, as we discuss in Appendix B.

We assume that the explicit training audience for such an enhanced WFX would remain the types of F/MF brigade headquarters that currently participate, given constraints on MCTP capacity. Nonetheless, at least some other brigade types might derive substantial training value from supporting such a large-scale FTX, even if they are not formal training audiences.

Recommendation Six

Prioritize unit participation in certain key joint and multinational exercises based on planned TPFDD alignments (if not already doing so). Existing joint and multinational exercises can provide substantial training value. For participation in such exercises, the Army should consider prioritizing units identified for OPLAN sourcing. In particular, this should apply to the types of F/MF brigade headquarters that are not included as WFX training audiences.

Recommendation Seven

The Army should provide F/MF brigade headquarters with EXEVALs as provided for in Army training doctrine; the Army should consider whether it is feasible and desirable to associate EXEVALs with enhanced WFXs (for brigades that participate in such exercises); and the Army should consider the advisability of resourcing such EXEVALs, partly using a permanent generating force organization to provide OC/Ts. Army doctrine states that all units in the Army must undergo an EXEVAL to validate training proficiency. Most of the brigade headquarters we spoke with stated that they would like to receive a formal EXEVAL, but none reported having received one. We recommend that the Army provide F/MF brigade headquarters with EXEVALs as provided for in Army training doctrine, or change its doctrine if it determines EXEVALs are not warranted or feasible at the brigade level.

The Army should also consider whether it is feasible and desirable to associate EXEVALs with enhanced WFXs (for brigades that participate in such exercises). This recommendation is premised on the notion that enhanced WFXs can and should be constructed to provide a venue appropriate for an EXEVAL—i.e., that providing an EXEVAL does not conflict with the notion of the WFX as a training event that provides units with the opportunity to fail and to learn from those failures. The Army should also determine which higher headquarters should conduct the EXEVAL (e.g., whether the EXEVAL should be conducted by a unit's higher headquarters in the TPFDD task organization, if this differs from the higher headquarters that exercises peacetime ADCON over the unit). Moreover, in the case of the SUST BDE, it would appear more suitable to have an ESC conduct the EXEVAL, even if the division to which the SUST BDE is nominally attached in peacetime happens to also be included in the TPFDD. Finally, the Army should consider whether it is feasible and valuable to resource a permanent generating force organization to provide OC/Ts for brigade-level EXEVALs, promote standardization, and better ensure that qualified OC/Ts are available to support EXEVALs.

Doctrine and METLs for Select F/MF Brigade Headquarters

This appendix first presents an overview of Army command and support relationships as background information. We then describe the doctrinal organizations, missions, and command and support relationships for the types of F/MF brigade headquarters that we examined.[1] For each type of brigade headquarters, we first describe the unit's designed mission and capabilities. We then list the unit's standardized and approved METL (METs are shown in bold font; SCTs are shown in regular font). Finally, we summarize functional doctrine for the unit type with regard to organization and to command and support relationships for LSCO, including input from stakeholder interviews as applicable.

Information on the brigade headquarters' mission and capabilities is taken verbatim from the unit's table of organization and equipment. These tables are available from FMSWeb, which is owned by the U.S. Army Force Management Support Agency.[2]

Information on METLs—both the unit's mission and the listing of METs and SCTs—is taken verbatim from the METL Viewer application on the Army Training Network website.[3]

Army Command and Support Relationships

Before turning to specific unit types, we first summarize Army command and support relationships (which are similar but not identical to joint command authorities and relationships). See Appendix A of FM 3-0, *Operations*, for more information.[4]

Command Relationships

Per FM 3-0,

> Army command relationships define superior and subordinate relationships between unit commanders. . . . Army command relationships identify the degree of control of the gaining Army commander. The type of command relationship often relates to the expected lon-

---

[1] In some instances, we have spelled in brackets the term to which an acronym refers, if the acronym was used by itself and not used elsewhere in the report. In cases where the original wording included both a term and its associated acronym, we deleted the acronym if it was not used elsewhere in the report.

[2] U.S. Army Directorate of Force Management, *FMSWeb*, database, undated.

[3] Army Training Network, undated.

[4] FM 3-0, *Operations*, Washington, D.C.: Headquarters, Department of the Army, October 6, 2017.

gevity of the relationship between the headquarters involved, and it quickly identifies the degree of support that the gaining and losing Army commanders provide.[5]

Army command relationships include the following:

- **Organic.** "Organic forces are those assigned to and forming an essential part of a military organization as listed in its table of organization. . . . If temporarily task-organized with another headquarters, organic units return to the control of their organic headquarters after completing the mission. To illustrate, within a brigade combat team (BCT), the entire brigade is organic. In contrast, within most functional and multifunctional brigades, there is a 'base' of organic battalions and companies and a variable mix of assigned and attached battalions and companies."[6]
- **Assigned.** "Assign is to place units or personnel in an organization where such placement is relatively permanent, and/or where such organization controls and administers the units or personnel for the primary function, or greater portion of the functions, of the unit or personnel. . . . Unless specifically stated, this relationship includes [ADCON]."[7]
- **Attached.** "Attach is the placement of units or personnel in an organization where such placement is relatively temporary. . . . A unit may be temporarily placed into an organization for the purpose of conducting a specific operation of short duration. Attached units return to their parent headquarters (assigned or organic) when the reason for the attachment ends. The Army headquarters that receives another Army unit through assignment or attachment assumes responsibility for the ADCON requirements, and particularly sustainment, that normally extend down to that echelon, unless modified by directives."[8]
- **OPCON.** "Operational control is the authority to perform those functions of command over subordinate forces involving organizing and employing commands and forces, assigning tasks, designating objectives, and giving authoritative direction necessary to accomplish the mission. . . . OPCON normally includes authority over all aspects of operations and joint training necessary to accomplish missions. It does not include directive authority for logistics or matters of administration, discipline, internal organization, or unit training."[9]
- **TACON.** "Tactical control is the authority over forces that is limited to the detailed direction and control of movements or maneuvers within the operational area necessary to accomplish missions or tasks assigned. . . . TACON is inherent in OPCON . . . TACON provides sufficient authority for controlling and directing the application of force or tactical use of combat support assets within the assigned mission or task. TACON does not provide organizational authority or authoritative direction for administrative and logistic support; the commander of the parent unit continues to exercise these authorities unless otherwise specified in the establishing directive."[10]

---

[5]   FM 3-0, 2017, p. A-3.

[6]   FM 3-0, 2017, p. A-4.

[7]   FM 3-0, 2017, p. A-5.

[8]   FM 3-0, 2017, p. A-5.

[9]   FM 3-0, 2017, p. A-1.

[10]   FM 3-0, 2017, p. A-2.

Support Relationships
According to FM 3-0:

> Army support relationships allow supporting commanders to employ their units' capabilities to achieve results required by supported commanders. Support relationships are graduated from an exclusive supported and supporting relationship between two units—as in direct support—to a broad level of support extended to all units under the control of the higher headquarters—as in general support. Support relationships do not alter [ADCON]. Commanders specify and change support relationships through task organization.[11]

- **Direct Support.** "Direct support is a support relationship requiring a force to support another specific force and authorizing it to answer directly to the supported force's request for assistance. A unit assigned a direct support relationship retains its command relationship with its parent unit, but it is positioned by and has priorities of support established by the supported unit." [12]
- **General Support.** "General support is that support which is given to the supported force as a whole. It is not given to any particular subdivision of the force. Units assigned a general support relationship are positioned and have priorities established by their parent unit." [13]
- **Reinforcing.** "Reinforcing is a support relationship requiring a force to support another supporting unit. Only like units (for example, artillery to artillery) can be given a reinforcing mission. A unit assigned a reinforcing support relationship retains its command relationship with its parent unit, but it is positioned by the reinforced unit. A unit that is reinforcing has priorities of support established by the reinforced unit, then the parent unit." [14]
- **General Support-Reinforcing.** "General support-reinforcing is a support relationship assigned to a unit to support the force as a whole and to reinforce another similar-type unit. . . . A unit assigned a general support-reinforcing support relationship is positioned and has its priorities established by its parent unit and secondly by the reinforced unit." [15]

CAB Headquarters

Mission
According to FM 3-0: "To provide command, control, staff planning, and supervision of Combat Aviation Brigade operations."[16]

Capabilities
"This unit provides:

---

[11] FM 3-0, 2017, p. A-6.

[12] FM 3-0, 2017, p. A-6.

[13] FM 3-0, 2017, p. A-6.

[14] FM 3-0, 2017, p. A-6.

[15] FM 3-0, 2017, pp. A-6–A-7.

[16] Army Training Network, undated.

- Command, control, supervision, and staff planning of brigade operations and operational control of additional air and ground maneuver elements. Liaison with higher and adjacent headquarters and supported and supporting elements.
- Unit level personnel service support, including administrative support, legal, force health protection (including flight surgeon support), and religious support for the Brigade [headquarters and headquarters company, or HHC].
- Unit level logistical support, including supply and communications maintenance (except [communications security, or COMSEC]) for the Brigade HHC.
- Coordination of Airspace Command and Control (AC2) requirements for aviation division assets.
- Equipment and unit level support to the U.S. Air Force Weather Team.
- Intelligence exploitation of sensor data generated by CAB aircraft when augmented by force provided Processing, Exploitation and Dissemination (PED) Platoon(s) from the Expeditionary Military Intelligence Brigade (E-MIB)."[17]

METLs
- **01-BDE-6114 Conduct Reconnaissance Operations**
  - 01-BDE-0016 Integrate Aircraft Survivability Measures into Mission Planning
  - 01-BDE-2300 Perform Information Collection
  - 01-BDE-6110 Integrate Airspace Control into Aviation Mission Planning
  - 01-BDE-8117 Coordinate Aviation Operations
  - 71-BDE-5100 Conduct the Mission Command Operations Process for Brigades
- **01-BDE-6115 Conduct Security Operations**
  - 01-BDE-0016 Integrate Aircraft Survivability Measures into Mission Planning
  - 01-BDE-2300 Perform Information Collection
  - 01-BDE-6110 Integrate Airspace Control into Aviation Mission Planning
  - 01-BDE-8117 Coordinate Aviation Operations
  - 71-BDE-5100 Conduct the Mission Command Operations Process for Brigades
- **01-BDE-6116 Conduct Air Movement Operations**
  - 01-BDE-0016 Integrate Aircraft Survivability Measures into Mission Planning
  - 01-BDE-2300 Perform Information Collection
  - 01-BDE-6110 Integrate Airspace Control into Aviation Mission Planning
  - 01-BDE-6119 Conduct Airfield Management Responsibilities (Brigade, Aviation)
  - 71-BDE-5100 Conduct the Mission Command Operations Process for Brigades
- **01-BDE-6117 Conduct Aeromedical Evacuation Operations**
  - 01-BDE-0448 Coordinate with the Joint Medical Structure for Aeromedical Evacuation
  - 08-BDE-1824 Manage Medical Evacuation Support (Air/Ground)
  - 08-BDE-1850 Coordinate Army Aeromedical Evacuation Support to Theater Health Service Support Operations
  - 08-BDE-9009 Manage Aeromedical Evacuation Support Operations
  - 71-BDE-5100 Conduct the Mission Command Operations Process for Brigades
- **01-BDE-6118 Conduct Offensive Operations**

---

[17] Army Training Network, undated.

    – 01-BDE-0016 Integrate Aircraft Survivability Measures into Mission Planning
    – 01-BDE-6110 Integrate Airspace Control into Aviation Mission Planning
    – 01-BDE-8117 Coordinate Aviation Operations
    – 71-BDE-5100 Conduct the Mission Command Operations Process for Brigades
- **55-BDE-4800 Conduct Expeditionary Deployment Operations at the Brigade Level**
    – 12-BDE-0004 Prepare Personnel for Deployment (S1)
    – 55-BDE-4801 Conduct Actions Associated with Force Projection at the Brigade Level
    – 55-BDE-4804 Conduct Deployment Activities at the Brigade Level
    – 71-BDE-5100 Conduct the Mission Command Operations Process for Brigades

Organic Structure

Like BCTs, CABs have a set of organic subordinate units that, for LSCO, would typically deploy and operate with the CAB headquarters. CABs include the following subordinate units:

- one attack/reconnaissance battalion
- one heavy attack/reconnaissance squadron
- one assault battalion
- one general support aviation battalion
- one aviation support battalion
- one Gray Eagle company.

The CAB headquarters can exercise mission command over up to two additional aviation battalions without staff augmentation, requires additional maintenance personnel and equipment.

Summary of Doctrine

Key branch-specific doctrinal publications include:

- FM 3-04, *Army Aviation*
- ATP 3-04.1, *Aviation Tactical Employment*.[18]

In LSCO, a CAB would either deploy with the division headquarters to which it is attached while at home station or be attached to a different division headquarters. It could also be attached to a corps headquarters. Per our unit interviews, in addition to its higher headquarters, the CAB must integrate and synchronize with the FA BDE and/or DIVARTY for fire targeting and airspace deconfliction. It must also work with the ADA brigade to a degree (but interviews indicated that the division headquarters is the primary link to the ADA brigade).
Per FM 3-0:

> The CAB's attack, reconnaissance, utility, and cargo aircraft may maneuver independently under corps or division control in the echelon deep area or within an assigned [area of operations]. Alternatively, the CAB's attack, reconnaissance, utility, and cargo assets may be under OPCON, TACON, general support, or direct support to another brigade as situationally appropriate. Furthermore, a CAB may receive OPCON of ground maneuver

---

[18] FM 3-04, 2015; ATP 3-04.1, *Aviation Tactical Employment*, Washington, D.C.: Headquarters, Department of the Army, April 13, 2016.

forces to conduct security or reconnaissance of the corps' or division's flanks or front or to accomplish other economy of force missions."[19]

Per our unit interviews, the CAB's subordinate assets would typically have support relationships with a BCT or another brigade type, rather than OPCON or TACON. However, per FM 3-04, a CAB "may place a company size unit OPCON to a ground force for a specific mission requirement, usually of limited and short duration where no enduring sustainment is required."[20]

One area where coordination with UAPs can be necessary involves the JAAT. Per ATP 3-04.1:

> The JAAT includes a combination of attack and/or scout [rotary-wing] aircraft and fixed-wing (FW) [close air support] aircraft operating together to locate and attack high-payoff targets and other targets of opportunity. A JAAT normally operates as a coordinated effort supported by fire support, ADA, naval surface fire support (NSFS), intelligence, surveillance, and reconnaissance systems, electronic warfare (EW) systems, and ground maneuver forces against enemy forces. [21]

FA BDE Headquarters

Mission
"Plan, prepare, execute and assess combined arms operations to provide close support and precision strike for the Corps employing Joint and organic fires and capabilities to achieve distribution effects in support of commanders' operational and tactical objectives."[22]

Capabilities
"This unit provides:

- Mission command of fires forces with a command authority, including recommending fires command and support relationships and positioning of Assigned/Attached/OPCON fires units.
- Integration of Army, Joint, and UAP tactical fires in support of distributed operations at the Corps level.
- Synchronization of counterfire operations and radar employment in the Corps area of operations.
- Establish common survey and meteorological data across the Corps area of operations.
- Conduct training, certification and mentoring of fires forces in the Corps (includes [echelon above brigade field artillery battalions] and Assigned/Attached/OPCON [field artillery battalions and field artillery brigades]).

---

[19] FM 3-0, 2017, p. 2-16.

[20] FM 3-04, 2015, p. 2-16.

[21] ATP 3-04.1, 2016, p. 2-41.

[22] Army Training Network, undated.

- Provide oversight of fires training and certifications in the Corps (Howitzer sections, Fire Direction Centers/Platoon Operations Centers, Fire Support Team, [Joint Forward Observer], Artillery leader, Radar section, Survey section.
- Conduct unit certifications (Artillery tables) at Battery and Battalion level for [echelon above brigade] Field Artillery Battalions."[23]

Organic Structure

FA BDE headquarters do not have a full set of organic subordinate units. FA BDEs each have an organic signal company and a brigade support battalion headquarters and service company. In operations, the FA BDE will be task organized with a combination of one to five rocket/missile and/or cannon field artillery battalions, and potentially other enablers.

METL
- **06-BDE-1084 Synchronize Fire Support**
  - 06-BDE-1035 Implement the Fire Support Plan
  - 06-BDE-1110 Conduct Field Artillery Liaison
  - 06-BDE-1118 Conduct Fire Support Planning
  - 06-BDE-2022 Evaluate the Fire Support Threat
  - 06-BDE-5059 Coordinate Target Attack
  - 17-BN-0308 Synchronize Close Air Support–Battalion
  - 71-BDE-5100 Conduct the Mission Command Operations Process for Brigades
- **06-BDE-5053 Control Field Artillery Fire Missions**
  - 06-BDE-1023 Coordinate a Field Artillery Raid
  - 06-BDE-6055 Process a Precision Fire Mission
  - 71-BDE-5100 Conduct the Mission Command Operations Process for Brigades
- **06-BDE-5060 Conduct Counterfire Operations**
  - 06-BDE-2005 Prepare the Information Collection Plan
  - 06-BDE-2035 Process Counterfire Target Information
  - 06-BDE-5074 Analyze Targets
  - 71-BDE-5100 Conduct the Mission Command Operations Process for Brigades
- **06-BDE-5066 Employ Fires**
  - 06-BDE-2011 Request Battle Damage Assessment
  - 06-BDE-6011 Conduct Fires
  - 71-BDE-5100 Conduct the Mission Command Operations Process for Brigades
- **06-BDE-5431 Execute Targeting Process**
  - 06-BDE-2005 Prepare the Information Collection Plan
  - 06-BDE-5078 Prepare Schedule of Fires
  - 06-BDE-5435 Provide Input to the Targeting Process
  - 06-BDE-6053 Conduct Precision Targeting
  - 06-BDE-6062 Conduct Joint Targeting
  - 71-BDE-5100 Conduct the Mission Command Operations Process for Brigades
- **06-BDE-6061 Integrate Sensors through Targeting**
  - 06-BDE-2006 Direct the Employment of Field Artillery Acquisition Assets

---

[23] Army Training Network, undated.

- 06-BDE-5437 Coordinate Target Acquisition and Counterfire
- 71-BDE-5100 Conduct the Mission Command Operations Process for Brigades
- **55-BDE-4800 Conduct Expeditionary Deployment Operations at the Brigade Level**
  - 12-BDE-0004 Prepare Personnel for Deployment (S1)
  - 55-BDE-4802 Conduct Home Station Mobilization Activities at the Brigade Level
  - 55-BDE-4804 Conduct Deployment Activities at the Brigade Level
  - 55-BDE-4850 Direct Deployment Alert and Recall at the Brigade Level
  - 55-BDE-4873 Plan Deployment at the Brigade Level
  - 55-EAC-4862 Coordinate Onward Movement
  - 71-BDE-5100 Conduct the Mission Command Operations Process for Brigades

Summary of Doctrine

Key branch-specific doctrinal publications include the following:

- ADRP 3-09, *Fires* (2013)
- FM 3-09, *Field Artillery Operations and Fire Support* (2014)
- ATP 3-09.24, *Techniques for the Fires Brigade*.[24]

The above three doctrinal publications have not yet been updated to distinguish between corps-level FA BDEs and DIVARTYs.[25] They instead refer to fires brigade headquarters that could support either division or corps headquarters. Current FA BDEs primarily support corps headquarters, while DIVARTYs are division-level assets.

In LSCO, an FA BDE would either deploy with the corps headquarters to which it is attached while at home station or be attached to a different corps headquarters. Per our interview with the Fires CoE interview, the FA BDE must also integrate and synchronize with the CAB and ADA brigade, as well as with joint and multinational forces.

Per FM 3-0:

A field artillery brigade's primary task is conducting corps-level strike operations. It is capable of employing Army fires and incorporating electronic warfare (EW). In addition, a brigade can request joint fires and coordinate with airspace control elements. The field artillery brigade can detect and attack targets using a mix of its organic target acquisition and fires capabilities, a supported division's information collection capabilities, and access to higher echelon headquarters information collection capabilities provided by the intelligence enterprise. Field artillery brigades are typically the force field artillery headquarters for the formation to which they are aligned. The field artillery brigade is capable of providing and coordinating joint lethal and nonlethal effects.[26]

In terms of subordinates, the FA BDE will be task organized to provide mission command over a combination of one to five field artillery battalions and possibly additional surveillance, reconnaissance, target acquisition, and other fire support assets. The FA BDE's sup-

---

[24] ATP 3-09.24, *Techniques for the Fires Brigade*, Washington, D.C.: Headquarters, Department of the Army, November 21, 2012.

[25] DIVARTYs are addressed in ATP 3-09.90, *Division Artillery Operations and Fire Support for the Division*, Washington, D.C.: Headquarters, Department of the Army, October 12, 2017.

[26] FM 3-0, 2017, p. 2-16.

porting assets might also include ground reconnaissance and surveillance, manned aviation, and unmanned aircraft assets as directed by higher echelons.[27] Subordinate elements might be attached or under the FA BDE's OPCON or TACON.

Field artillery doctrine stresses "maximum feasible centralized control . . . consistent with the fire support capabilities and requirements of the overall mission."[28] Field artillery battalions under the FA BDE will have support relationships—e.g., direct support, reinforcing, general support—rather than being placed under the command of maneuver forces. General and general support-reinforcing relationships provide relatively more centralized control for the FA BDE, whereas reinforcing and direct support relationships provide relatively less centralized control, although all support relationships leave the FA BDE in command of its subordinate battalions. Per FM 3-09, for example:

> Since a movement to contact involves an unclear or uncertain situation, the [FA BDE] maintains centralized control [i.e., field artillery battalions will have general support or general support reinforcing relationships with the maneuver force] . . . enabling a coordinated response to a rapidly developing situation... Transitioning [FA BDE] cannon field artillery battalions from general support (GS) or general support reinforcing (GSR) to reinforcing (R) once the situation is developed and the BCTs begin to conduct follow-on offensive or defensive tasks.

> During the attack, less centralized control of fires is used because the [FA BDE]-supported maneuver force will have the initiative.

> During exploitation and pursuit it is important to have decentralized execution authority and support relationships. For example . . . [FA BDE] cannon field artillery battalions may have a reinforcing (R) support relationship to BCT field artillery battalions.

> In defensive tasks, the supported commander normally directs more centralized control of all artillery assets, to ensure they are immediately responsive to the supported headquarters.[29]

The FA BDE might support JAAT operations. Moreover,

> Multinational combinations are common and interagency and intergovernmental combinations are occurring more frequently. None of these (joint, interagency, intergovernmental, or multinational) combinations occurs successfully in operations without planning and preparation. Careful training and exchange of liaison at every level are necessary for successful operations.[30]

---

[27] FM 3-09, 2014, p. 1-36.

[28] ADRP 3-09, 2013, p. 1-14.

[29] FM 3-09, 2014, pp. 1-7, 1-8, 1-11, 1-16.

[30] ATP 3-09.24, 2012.

SUST BDE Headquarters

Mission
"To provide mission command for attached units, plan, coordinate and synchronize sustainment operations."

Capabilities
"This unit provides:
- Mission command of attached units.
- Planning, coordinating, and synchronizing sustainment operations.
- Administrative, communication equipment, [chemical, biological, radiological, nuclear, and high explosive] defense, electronic warfare, and sustainment automation management support for attached units.
- Provides field maintenance for the [HHC] and [special troops battalion] Sustainment Brigade and Brigade Signal Company."

Organic Structure
SUST BDE headquarters do not have a full set of organic subordinate units. All Regular Army SUST BDE headquarters have an organic signal company, but no other organic elements. In operations, the SUST BDE will be task organized with a combination of three to seven subordinate sustainment battalions.

METL
- **12-EAC-1228 Coordinate Human Resources Support During Offense, Defense, Stability, and Defense Support of Civil Authorities Operations (Sustainment Brigade [SB]–Human Resources Operations Branch [HROB])**
  - 12-BDE-0009 Process Replacements (S1)
  - 12-BDE-0036 Conduct Personnel Accountability (S1)
  - 12-BDE-0037 Conduct Strength Reporting (S1)
  - 12-EAC-1216 Monitor Transient Personnel Activities (SB-HROB)
  - 12-EAC-1232 Manage Casualty Reporting (HROB)
  - 12-EAC-1254 Manage Postal Services (HROB)
  - 71-BDE-5100 Conduct the Mission Command Operations Process for Brigades
- **14-EAC-8025 Provide Funding Support to Financial Management Elements During Offense, Defense, Stability, and Defense Support of Civil Authorities Operations (Financial Management Support Unit [FMSU])**
  - 14-EAC-8002 Perform Disbursing Operations (FMSU)
  - 14-EAC-8004 Conduct Commercial Vendor Pay Operations (FMSU)
  - 14-EAC-8008 Provide Military Pay Support (FMSU)
  - 14-EAC-8017 Conduct Audit Readiness Compliance Operations (FMSU)
  - 14-EAC-8018 Provide Technical Guidance to Financial Management Elements (Financial Management Support Center)
  - 14-EAC-8019 Maintain Financial Management Systems (FMSU)
  - 71-BDE-5100 Conduct the Mission Command Operations Process for Brigades
- **55-BDE-4800 Conduct Expeditionary Deployment Operations at the Brigade Level**
  - 55-BDE-4804 Conduct Deployment Activities at the Brigade Level

- 55-BDE-4878 Conduct Redeployment Activities at the Brigade Level
- 55-EAC-4864 Conduct Home Station Rear Detachment Operations at Echelon Above Corps
- 63-BDE-0055 Integrate Operational Contract Support into Mission Command
- 71-BDE-5100 Conduct the Mission Command Operations Process for Brigades
- **63-BDE-0002 Manage Logistics Support in the Operational Area During Offense, Defense, Stability, and Defense Support of Civil Authorities Operations**
  - 08-BDE-9008 Conduct Surgeon Section Activities–Surgeon
  - 63-BDE-0003 Develop Logistics Estimate
  - 63-BDE-0047 Integrate Joint, Interagency, Intergovernmental, and Multinational Sustainment
  - 63-BDE-4061 Provide Liaison Support
  - 63-EAC-0050 Conducts Boards, Bureaus, Centers, Cells, and Working Groups
  - 63-EAC-2917 Manage Distribution Operations
  - 71-BDE-5100 Conduct the Mission Command Operations Process for Brigades
- **63-BDE-0727 Conduct Actions Associated with Area Defense During Offensive, Defensive, Stability, and Defense Support of Civil Authorities Operations**
  - 03-SEC-9007 Coordinate [Chemical, Biological, Radiological, and Nuclear] Protection
  - 55-BN-0055 Plan Tactical Convoy During Offense, Defense, Stability and Defense Support of Civil Authorities Operations
  - 63-EAC-2613 Coordinate Security Activities
  - 63-EAC-4013 Plan Base and Base Cluster Operations
  - 71-BDE-5100 Conduct the Mission Command Operations Process for Brigades
- **63-BDE-4060 Coordinate Distribution Operations During Offense, Defense, Stability, and Defense Support of Civil Authorities Operations**
  - 10-6-4024 Coordinate Bulk Petroleum Distribution
  - 10-9-2415 Distribute Bulk Water During Offense, Defense, Stability, and Defense Support of Civil Authorities Operations
  - 10-BDE-4031 Plan Field Services Support
  - 63-BDE-2401 Manage Supply Support
  - 63-BDE-2453 Coordinate Transportation Support
  - 63-BDE-4877 Provide Sustainment Support During Offense, Defense, Stability and Defense Support of Civil Authorities Operations
  - 71-BDE-5100 Conduct the Mission Command Operations Process for Brigades
- **63-CMD-0027 Conduct Reception, Staging, and Onward Movement During Offense, Defense, Stability, and Defense Support of Civil Authorities Operations**
  - 55-EAC-0003 Manage Transportation Operations in Support of Reception, Staging, and Onward Movement During Offense, Defense, Stability, and Defense Support of Civil Authorities operations
  - 63-BDE-1201 Conduct Staging and Marshalling Area Activities
  - 63-BDE-2457 Coordinate Onward Movement Operations
  - 63-BDE-2459 Coordinate Reception of Forces at Sea/Aerial Port of Debarkation/Embarkation
  - 63-BDE-2463 Manage Reception of Forces at Ports of Debarkation
  - 63-BDE-2477 Provide Life Support at Terminals and Reception Nodes
  - 71-BDE-5100 Conduct the Mission Command Operations Process for Brigades

Summary of Doctrine

Key branch-specific doctrinal publications include:

- ADRP 4-0, *Sustainment* (2012)
- ATP 4-93, *Sustainment Brigade.*[31]

In LSCO, an SUST BDE would normally be attached to an ESC or TSC. Per our interview with CASCOM, this command relationship enables sustainment chain of command to use force as needed for any specific part of the battlefield. According to FM 3-0, a SUST BDE

> normally attaches sustainment brigades to an ESC with a general or area support relationship to a corps or division headquarters. However, during high tempo large-scale combat operations, a sustainment brigade may be placed OPCON to a corps or division based on the mission and operational variables. . . .
>
> A sustainment brigade is capable of providing general support to one or more divisions, BCTs, multifunctional and functional brigades, ancillary units, and unified action partners. This is in addition to supporting the corps headquarters and headquarters battalion and other units operating in its assigned area. . . .
>
> The brigade focuses on management and distribution of supplies, field services, human resources support, execution of financial management support, and allocation of field echelon maintenance in an assigned area.[32]

In terms of subordinates, the SUST BDE has no organic subordinates other than its signal company. Although it might deploy with subordinate battalions that are under its habitual ADCON at home station, it might not—its deployed task organization might include only attached subordinate battalions with which it has no habitual relationships. The SUST BDE will command a combination of three to seven multifunctional CSSBs and functional sustainment battalions (e.g., transportation, ordnance, quartermaster). The SUST BDE's task organization changes based on changing mission requirements. The sustainment brigade and its attached units will normally have a general support relationship with supported organizations."[33]

According to ATP 4-93:

> (T)here are instances when the sustainment brigade communicates and coordinates directly with unified action partners' representatives to synchronize and integrate support. . . . This coordination will be more common for a sustainment brigade supporting operational level forces, a theater opening mission or a largely contracted mission. Sustainment brigade commanders and staff must be familiar with U.S. governmental partners and understand what each partner provides to support ASCC objectives.

However, our CASCOM interview suggested that challenges related to integrating with UAPs in an LSCO were likely to fall more to the ESC and/or TSC than to SUST BDEs.

---

[31] ADRP 4-0, *Sustainment*, Washington, D.C.: Headquarters, Department of the Army, July 13, 2012; ATP 4-93, *Sustainment Brigade*, Washington, D.C.: Headquarters, Department of the Army, April 11, 2016.

[32] FM 3-0, 2017, p. 2-17.

[33] ATP 4-93, 2016, p. 1-1.

EN BDE Headquarters

Mission
"Provides mission command for all assigned or attached engineer teams, companies and battalions echelons above brigades [sic]. When directed, it may also provide mission command for engineers from other Services and multinational forces. Its mission focus is on operational-level engineer support across all three of the engineer disciplines and may serve as the Theater Engineer Command for a Theater Army, land component headquarters, or a [joint task force]."[34]

Capabilities
"This unit provides:

- Planning, supervision, and coordination for combat engineer support, construction, facility rehabilitation, unit allocation, resource management, river crossings, barrier placement, countermine and counter-obstacle operations.
- Supervision of contract construction, labor, and indigenous personnel.
- Planning and supervision for terrain intelligence and topographic operations."[35]

Organic Structure
EN BDE headquarters have no organic subordinate units. In operations, the EN BDE headquarters will be task organized with a combination of two to five subordinate engineer battalions.

METL
- **05-BDE-0096 Conduct General Engineering Operations**
  - 05-BDE-0082 Provide Engineer Support for Real Property Maintenance
  - 05-BDE-0084 Coordinate Engineer Support with the Host Nation/Coalition Representative
  - 05-BDE-0714 Conduct an Infrastructure Assessment
  - 05-BDE-0715 Coordinate Construction Operations
  - 71-BDE-5100 Conduct the Mission Command Operations Process for Brigades
- **05-BDE-0098 Conduct Countermobility Operations**
  - 05-BDE-2000 Prepare an Obstacle Plan
  - 71-BDE-5100 Conduct the Mission Command Operations Process for Brigades
- **05-BDE-0099 Conduct Survivability Operations**
  - 05-BDE-0080 Recommend the Employment Priority of Engineer Assets
  - 05-BDE-0812 Conduct Equipment Support Missions
  - 71-BDE-5100 Conduct the Mission Command Operations Process for Brigades
- **05-BDE-1007 Conduct Mobility Operations**
  - 05-BDE-0024 Integrate Engineer Information Requirements into the Intelligence, Surveillance, and Reconnaissance Plan
  - 05-BDE-0073 Provide Engineer Support to Breaching Operations
  - 05-BDE-0640 Conduct Gap Crossing Operations
  - 05-BDE-1006 Conduct Counter-[Improvised Explosive Device] Operations

---

[34] Army Training Network, undated.

[35] Army Training Network, undated.

- 05-BDE-6003 Provide Geospatial Support
- 71-BDE-5100 Conduct the Mission Command Operations Process for Brigades
- **55-BDE-4800 Conduct Expeditionary Deployment Operations at the Brigade Level**
  - 12-BDE-0004 Prepare Personnel for Deployment (S1)
  - 55-EAC-4804 Conduct Deployment Activities at Echelon Above Corp
  - 55-EAC-4850 Direct Deployment Alert and Recall at Echelons Above Corp
  - 55-EAC-4864 Conduct Home Station Rear Detachment Operations at Echelon Above Corps
  - 55-EAC-4873 Plan Deployment at the Echelons Above Corp Level
  - 71-BDE-5100 Conduct the Mission Command Operations Process for Brigades

Summary of Doctrine

Key branch-specific doctrinal publications include:

- FM 3-34, *Engineer Operations*
- ATP 3-34.23, *Engineer Operations: Echelons Above Brigade Combat Team.*[36]

In LSCO, an EN BDE would either deploy with the corps headquarters to which it is attached while at home station or be attached to a different corps headquarters. According to FM 3-0:

> An engineer brigade integrates and synchronizes engineer capabilities across a corps [area of operations] and reinforces subordinate corps units in the execution of engineer tasks by allocating mission-tailored engineer forces. . . . Engineer tasks alter terrain to overcome obstacles (including gaps), create, maintain, and improve lines of communication, create fighting positions, improve protective positions, and build structures and facilities (including base camps, aerial ports, seaports, utilities, and buildings).[37]

At division level, engineer assets not organic to BCTs are typically controlled by the MEB headquarters. Per ATP 3-34.23, an EN BDE headquarters might be attached to a division headquarters for certain missions that exceed the MEB's ability to provide mission command.[38] Per our unit interviews, this could arise in relation to a major breaching or wet gap crossing operation.

In terms of subordinates, the EN BDE has no organic subordinate elements of any type. It might deploy with subordinate battalions that are under its habitual ADCON at home station, but it might not—its deployed task organization might include only attached subordinate battalions with which it has no habitual relationships. according to FM 3-0, "An engineer brigade can control up to five mission-tailored engineer battalions . . . (that) have capabilities

---

[36] FM 3-34, *Engineer Operations*, Washington, D.C.: Headquarters, Department of the Army, April 2, 2014; ATP 3-34.23, *Engineer Operations: Echelons Above Brigade Combat Team*, Washington, D.C.: Headquarters, Department of the Army, June 10, 2015.

[37] FM 3-0, 2017, p. 2-19.

[38] ATP 3-34.23, 2015, p. 5-8.

from any of the three engineer disciplines, combat engineering, general engineering, and geospatial engineering."[39]

Regarding command and support relationships for subordinate battalions, ATP 3-34.23 states:

> Experience has generally shown that command relationships work well in offensive operations, but that support relationships allow for the efficient use of high-demand, low-density engineering capabilities during defensive and stability operations. . . . For major combat operations, a significant portion of the tailored engineer force tends to have command relationships to maneuver commanders. The tailored engineer force will be pushed using command relationships in the task organization to tactical echelons for close support of combat operations. This will be true for some general engineering capabilities and for most combat engineering capabilities. Movement and maneuver requirements are not well defined at higher echelons and are more dynamic in combat operations. Tailored forces are pushed to subordinate echelons to address these requirements and add flexibility for those maneuver commanders to react to unforeseen challenges and opportunities. . . . OPCON is the most common command relationship for engineers during offensive operations. It provides the gaining commander with the flexibility of a command relationship without the burden of [ADCON] responsibilities. . . . (For defensive operations) [s]upport relationships for engineer units to maneuver commanders are preferred to maximize the application of limited assets.[40]

EN BDEs might need to coordinate with UAPs. According to FM 3-34, for example: "With augmentation, the engineer brigade may serve as a joint engineer headquarters and may be the senior engineer headquarters deployed in a joint operations area if full theater engineering command deployment is not required."[41] However, our interviews did not identify any specific significant challenges for LSCO related to coordination with UAPs.

TTSB Headquarters

Mission
"To provide mission command of assigned and attached units. Supervises the installation, operation and maintenance of from 60 to 180 command post communications systems in the DOD Information Network–Army."

Capabilities
"This unit provides:

- Command, control, staff planning and supervision of the Theater Tactical Signal Brigade.
- Planning, engineering, and control of the Theater Communications System.
- Coordination of the training, administration, and logistical support of assigned units."[42]

---

[39] FM 3-0, 2017, pp. 2-19, 3-15, 4-11.

[40] ATP 3-34.23, 2015, p. 2-11.

[41] FM 3-34, 2014, p. 1-10.

[42] Army Training Network, undated.

Organic Structure

TTSBs have no organic subordinate units. In operations, the TTSB will be task organized with up to five signal battalions and any other forces necessary.

METL

- **11-BDE-9000 Conduct Department of Defense Information Network Operations**
  - 11-CW-6530 Conduct Cybersecurity
  - 11-CW-7113 Conduct Information Dissemination Management/Content Staging Activities
  - 11-CW-7128 Install a Local Area Network
  - 11-CW-7166 Conduct Enterprise Management
  - 11-CW-7379 Conduct Spectrum Management Operations
  - 11-CW-8013 Operate a Combat Network Radio System
  - 71-BDE-5100 Conduct the Mission Command Operations Process for Brigades
- **11-BDE-9010 Plan Department of Defense Information Network Operations**
  - 11-CW-0732 Plan Cybersecurity
  - 11-CW-7217 Plan Enterprise Management
  - 11-CW-7400 Plan Information Dissemination Management/Content Staging Activities
  - 11-CW-8004 Plan a Combat Network Radio Network
  - 71-BDE-5100 Conduct the Mission Command Operations Process for Brigades
- **55-BDE-4800 Conduct Expeditionary Deployment Operations at the Brigade Level**
  - 55-BDE-4802 Conduct Home Station Mobilization Activities at the Brigade Level
  - 55-BDE-4804 Conduct Deployment Activities at the Brigade Level
  - 55-BDE-4863 Coordinate Rear Detachment Support at the Brigade Level
  - 55-BDE-4873 Plan Deployment at the Brigade Level
  - 55-BDE-4878 Conduct Redeployment Activities at the Brigade Level
  - 55-CO-4803 Perform Predeployment Training Activities
  - 71-BDE-5100 Conduct the Mission Command Operations Process for Brigades

Summary of Doctrine

The key branch-specific doctrinal publication is the following:

- FM 6-02, *Signal Support to Operations*.[43]

There is limited doctrine regarding the TTSB; for example, doctrine governing the TTSB in FM 6-02 is limited to about one paragraph. Moreover, what doctrine there is appears to be somewhat out of date—characterizing the TTSB as primarily a theater asset (as the name would imply). Per FM 6-02:

> The theater tactical signal brigade conducts mission command for assigned and attached units. The HQs supervises the installation, operation, and maintenance of communications signal nodes, and engineers and defends these nodes, in the theater communications system, excluding the division and corps systems. It provides theater-level planning and

---

[43] FM 6-02, *Signal Support to Operations*, Washington, D.C.: Headquarters, Department of the Army, January 22, 2014.

engineering for mission command networks and systems, and baseline services. It also supervises the installation, operation, and maintenance of nodal communications in support of the Theater Army, coalition, and augmentation to the corps, division, other government agencies, and nongovernment organizations.[44]

An interviewee with a TTSB stated, however, that current planning provides that a TTSB will more likely act as a corps signal brigade. This interviewee stated that, when the TTSBs were originally designed, it was assumed that corps G-6s would have the capabilities needed to set up a corps-level network. However, changes since have reduced the corps headquarters' capability in this regard. FM 3-0 (2017), appears to bear this out:

> A TTSB provides functional signal support for corps and division operations. TTSBs provision communications and information systems support to a theater army headquarters, their subordinate units, and as required, to joint, inter-organizational, and multinational partners throughout the area of responsibility. The TTSB and its subordinate units install, operate, maintain, and defend the Department of Defense information network-Army (DODIN-A). Each TTSB leverages the extension and reachback capabilities to provide joint communications and information systems services to the [geographic combatant commander] and subordinate commanders to conduct mission command.[45]

In terms of subordinates, the TTSB has no organic subordinate elements. It might deploy with subordinate battalions that are under its habitual ADCON at home station, but it might not—its deployed task organization might include only attached subordinate battalions with which it has no habitual relationships. The TTSB will command up to five ESBs, and any other attached forces as necessary based on the direction of higher headquarters.

E-MIB Headquarters

Mission
"The E-MIB HQs exercises mission command over all assigned and attached elements supporting the deployed corps/combined joint task force (CJTF). Provides reach back capability to the greater intelligence enterprise when deployed. Deploys [E-MIB] in support of division(s) and provides mission command of division MI forces as directed. The composition of the deployed formation(s) is dependent on the required capabilities for that specific operation. The company headquarters (HQ) provides mission command and limited unit level administrative and logistics support to the brigade HQ."[46]

Capabilities
"This unit provides:

- Planning, coordination, synchronization, integration and execution of the following intelligence functions: asset and mission management; COMSEC oversight; signal intel-

---

[44] FM 6-02, 2014, p. 2-9.

[45] FM 3-0, 2017, p. 2-20.

[46] Army Training Network, undated.

ligence collection and exploitation; imagery intelligence exploitation; geospatial intelligence analysis and products; electronic preparation of the battlefield; counterintelligence and human intelligence operations.

- Mission command of assigned and attached elements of the brigade.
- Unit level administration for the brigade elements.
- Unit level ministry support for the brigade elements.
- Management of the brigade property book."[47]

Organic Structure

Each E-MIB headquarters has two organic expeditionary MI battalions.

METL

- **34-BDE-3001 Direct Operational Intelligence Activities**
  - 34-BN-0018 Manage Interrogation and Debriefing Operations at a Detention Facility
  - 34-SEC-3171 Plan Integration into the Intelligence Architecture
  - 71-8-5126 Integrate New Units/Soldiers into the Force (Battalion–Corps)
  - 71-BDE-5100 Conduct the Mission Command Operations Process for Brigades
  - 71-JNT-2140 Designate Intelligence Resources for Allocation in the Joint Operations Area for Joint Task Force
  - 71-JNT-5200 Assess the Tactical Situation for Operations for a Joint Task Force
- **34-BDE-3002 Collect Relevant Information**
  - 34-CO-3003 Conduct Human Intelligence Collection
  - 34-CO-3004 Conduct Signals Intelligence Collection
  - 34-CO-3007 Process Signals and Geospatial Imagery Intelligence
  - 34-SEC-0823 Perform Multifunctional Team Missions
  - 71-BDE-5100 Conduct the Mission Command Operations Process for Brigades
- **34-BDE-3003 Distribute Operational Intelligence**
  - 34-SEC-1204 Access External Databases
  - 34-SEC-1205 Conduct Analysis with Theater and National Intelligence Agencies Through [Joint Worldwide Intelligence Communications System] and [Secret Internet Protocol Router Network]
  - 71-BDE-5100 Conduct the Mission Command Operations Process for Brigades
  - 71-CORP-5310 Manage Information and Data
  - 71-CORP-5330 Integrate Shared Understanding Through Knowledge Management
  - 71-JNT-2421 Provide Warning Intelligence in the Joint Operations Area for Joint Task Forces
- **55-BDE-4800 Conduct Expeditionary Deployment Operations at the Brigade Level**
  - 12-BDE-0004 Prepare Personnel for Deployment (S1)
  - 55-BDE-4801 Conduct Actions Associated with Force Projection at the Brigade Level
  - 55-BDE-4804 Conduct Deployment Activities at the Brigade Level
  - 55-BDE-4805 Conduct Predeployment Activities at the Brigade Level
  - 55-EAC-4850 Direct Deployment Alert and Recall at Echelons Above Corp
  - 55-EAC-4873 Plan Deployment at the Echelons Above Corp Level

---

[47] Army Training Network, undated.

– 71-BDE-5100 Conduct the Mission Command Operations Process for Brigades

Summary of Doctrine

Key branch-specific doctrinal publications include:

- ADRP 2-0, *Intelligence*
- ATP 2-19.3, *Corps and Division Intelligence Techniques*.[48]

Currently, doctrine specific to E-MIBs is limited. Only a few sentences in ADRP 2-0 refer explicitly to E-MIBs. ATP 2-19.3 has an appendix on E-MIBs, but with the caveat that "Army-level discussions concerning the role, organization, and capabilities of the E-MIB are ongoing. Upon approval of final decisions, E-MIB discussions will be updated."[49]

In LSCO, an E-MIB would either deploy with the corps headquarters to which it is attached while at home station or be attached to a different corps headquarters. According to FM 3-0:

> The Army designed [E-MIBs] to augment the corps and division capability to process, exploit, and disseminate national and joint force signals intelligence and geospatial intelligence. E-MIBs also provide counterintelligence, human intelligence, and ground-based signals intelligence collection to corps and division headquarters. The E-MIB also supports site exploitation operations. The E-MIB does not conduct reconnaissance. (The corps and division commanders task available maneuver forces to conduct reconnaissance.) The corps commander retains control of the E-MIB or task organizes elements of the E-MIB to divisions as required.[50]

Although an E-MIB headquarters has two organic expeditionary MI battalions, the battalions might be attached to division headquarters while the E-MIB headquarters operates in support of a corps. In any case, the E-MIB headquarters provides mission command for any of its organic assets that remain under its control, as well as any assets attached at the corps' direction, such as MI aerial reconnaissance battalions. The E-MIB might also need to integrate with the CAB regarding "mission coordination" of the CAB's Gray Eagle company for intelligence, surveillance, and reconnaissance collection.

Current doctrine does not highlight significant issues regarding E-MIBs role in coordinating with UAPs. Unit interviews suggested that communications could be an issue but did not otherwise highlight any major roles or specific challenges.

---

[48] ADRP 2-0, *Intelligence*, Change No. 2, Washington, D.C.: Headquarters, Department of the Army, September 4, 2018; ATP 2-19.3, *Corps and Division Intelligence Techniques*, Change No. 1, Washington, D.C.: Headquarters, Department of the Army, March 10, 2016.

[49] ATP 2-19.3, 2016, p. 1-6.

[50] FM 3-0, 2017, p. 2-19.

Large-Scale Training and Exercise Opportunities Prior to OIF

The purpose of this appendix is to describe the opportunities for participation in major training events and exercises that were available to at least certain types of supporting brigade headquarters (e.g., aviation, field artillery, engineer) from the latter portion of the Cold War through the start of OIF. We refer to *supporting brigade headquarters* rather than *F/MF brigade headquarters*, since the latter term is a product of Army modularity that occurred during the 2000s and was not in use prior to OIF. Unfortunately, published sources that discuss the role of supporting brigade headquarters in major training event and exercises are limited. Without archival research and interviews—which were beyond the scope of this effort—it was impossible to make a detailed assessment of the number and types of supporting brigade headquarters that participated in the various events over time nor to assess the quality of those experiences in meeting their training objectives and preparing them to conduct LSCO. Nonetheless, this section provides at least some useful background on how the training environment prior to OIF compares with today. We first discuss the Army's CTC program. We next discuss large-scale joint and combined exercises. Finally, we discuss training opportunities prior to the start of LSCO in ODS and OIF.[1]

The Army's CTC Program

The Army's CTC program formally began in 1987.[2] The four CTCs include the three dirt CTCs—NTC, JRTC, and JMRC (then referred to as the Combat Maneuver Training Center)—and the BCTP, which is today's MCTP. As it does today, the CTC program prior to OIF focused on maneuver brigades at the dirt CTCs, and on division and corps headquarters through BCTP.[3] As with today's MCTP, the BCTP prior to OIF focused on CPXs supported

---

[1]   We do not discuss home-station training in this appendix. Just as today, however, such training could provide important opportunities. These include both formal training activities and support to other units deploying for operations or exercises. For example, maneuver brigades going to the CTCs exercised the capabilities of various home-station support units, even if the latter did not deploy. Moreover, supporting brigade headquarters attached to divisions experienced regular mission command training opportunities when their subordinate battalions operated in direct support of the division's maneuver brigades. (In other words, maneuver brigade headquarters had no permanently attached units; the equivalent of the division's artillery brigade provided field artillery units, the division support command—the rough equivalent to an SB of today—provided sustainment units, the division aviation brigade provided aviation elements, etc.) As these examples suggest, opportunities for home station training could be greater for supporting brigades attached to divisions, as opposed to those attached at corps or theater level.

[2]   The NTC at Fort Irwin started operating in 1982.

[3]   As it does today, BCTP also provided brigade WFXs for ARNG BCTs.

by computer simulations and did not provide field training opportunities such as maneuver brigades received at the dirt CTCs.

Prior to OIF, however, supporting brigade headquarters were not a *formal* BCTP training audience—not even on a secondary basis. Although supporting brigade headquarters of various types participated in WFXs to support the training objectives of division and corps headquarters, BCTP was simply not organized and resourced to focus on supporting brigade headquarters as a training audience. In other words, supporting brigade headquarters that participated in WFXs received the opportunity to set up their command posts and to exercise their processes and systems. Moreover, prior to modularity, some types of brigade headquarters operated with the brigade commander and substantial portions of the staff embedded with the headquarters of the division or corps to which they were aligned, meaning that BCTP provided a significant experience in terms of integrating with these higher echelons. On the other hand, WFXs were simply not designed to support the training objectives of supporting brigade headquarters and supporting brigade headquarters received no dedicated support from BCTP's OC/Ts. BCTP only began including various types of F/MF brigade headquarters as a training audience starting in 2007.[4] In short, today's MCTP provides a more substantial training opportunity for participating F/MF brigade headquarters than BCTP provided to similar units prior to OIF.

Large-Scale Joint and Combined Exercises

Up until the early 1990s, the Army also participated in a number of large-scale joint and combined exercises. REFORGER in Europe and TEAM SPIRIT in South Korea were the largest.[5] In addition to other activities,[6] each included a field training component that simulated an LSCO environment at corps level or higher.[7] In some iterations, these field exercises included the movement of whole divisions or large portions thereof, including various supporting brigade headquarters commanding substantial subordinate formations.[8] In addition

---

[4]   BCTP, undated.

[5]   Although this section focuses on the largest exercises (which approximated division- or corps-level operations), other smaller, but still fairly large-scale, exercises, such as BRIGHT STAR in Egypt, provided useful training opportunities; many small-scale exercises and theater security cooperation events provided opportunities as well. In particular, the latter often provided opportunities for nondivisional brigades to exercise mission command, even if they did not involve large numbers of subordinate units or opportunities to integrate with parallel and higher echelons in the same way as the large-scale exercises discussed in this appendix.

[6]   Each also involved the deployment of U.S. joint forces from the United States to training with U.S. and allied units already in theater. This was done both to exercise deployment capabilities and to signal U.S. resolve.

[7]   For example, "REFORGER 88's main field exercise, CERTAIN CHALLENGE, executed during 12–22 September, encompassed a maneuver area of 16,000 square miles in the West German states of Bavaria, Baden Wuerttemberg, and Hesse along a 100-mile front, which ran from Heilbronn east to Bamberg. Pitting two corps against each other, the CERTAIN CHALLENGE scenario posed a conflict between two fictional countries, Northland and Southland. Southland forces moved into a disputed part of the maneuver area, and the Northland forces sought to drive them back. Rather than a NATO versus Warsaw Pact exercise, both corps commanders tested NATO doctrine and enjoyed independent control over their units" (Terrence J. Gough, *Department of the Army Historical Summary: Fiscal Year 1986*, Washington, D.C.: Center of Military History, U.S. Army, 1995, p. 48).

[8]   An officer from the 5th ID described the unit's experience in REFORGER 78: "The months of August to mid-October 1978 provided one of the most valuable training experiences in the Red Devils' recent history. . . . (Prior to the exercise) a

to REFORGER and TEAM SPIRIT, the Army also participated in somewhat comparable large-scale joint exercises in the U.S. desert southwest. The BORDER STAR and GALLANT EAGLE series are examples. During these exercises, key training events were distributed across large installations, such as Fort Bliss, WSMR, Fort Irwin, and the Marine Corps Air Ground Combat Center at Twenty-Nine Palms; in addition, air support was staged from a variety of air bases throughout the region, and, in some cases, road movements between installations were a dimension of the exercise.[9] We were unable to determine whether these latter exercises provided the same types of training opportunities for supporting brigade headquarters that appear evident in the REFORGER and TEAM SPIRIT series. At a minimum, they provide interesting examples of the joint force using multiple, large training complexes across the desert southwest to support large-scale FTXs.

These large-scale FTXs faced some constraints. For example, they generally did not match the tactical realism of a dirt CTC rotation for an individual maneuver brigade or battalion task force. In REFORGER, for example, in many cases, movements were road-bound—or at least limited in their cross-country dimension—to minimize property damage and other adverse

---

5th Division REFORGER planning group . . . choreographed the intricate logistical and administrative steps involved in moving 13,000 troops, 117 helicopters, 503 wheeled and tracked vehicles, and nearly 5,000 tons of other equipment from CONUS installations to Germany and back . . . During FTX Certain Shield, the division exercised control over its own 1st Brigade, its division artillery, and its combat support and combat service support units, as well as over other units, including the European-based 11th Armored Cavalry Regiment, the 4th Brigade of the 4th ID, the 41st Field Artillery Group, the 11th Aviation Group, a brigade task force of the 2d British Armored Division, a Belgian mechanized infantry battalion, and an infantry company from Luxemburg" (Charles L. Wascom, "5th Infantry Division," *Infantry: A Professional Journal for the Combined Arms Team*, Vol. 69, No. 3, May–June 1979, pp. 17–18).

[9] For example, BORDER STAR 85 was a joint readiness exercise sponsored by the U.S. Readiness Command that involved approximately 30,000 soldiers, airmen, and Marines.

> Held at Fort Bliss, Texas, White Sands Missile Range, New Mexico, and adjacent public lands, the exercise used opposing forces in a simulated combat environment to train, test, and evaluate commanders, staffs, and forces in joint operations. Participants included I Corps Headquarters, units of the 9th ID, the 3d Armored Cavalry Regiment, 9th and 12th Air Force units, and elements of the 4th Marine Air Wing. More than thirty National Guard units and over fifty Army Reserve units also participated (Karl E. Cocke, *Department of the Army Historical Summary: Fiscal Year 1985*, Washington, D.C.: Center of Military History, U.S. Army, 1995, p. 33).

Approximately 500 tracked vehicles and 2,500 wheeled vehicles participated, as well as about 300 helicopters and airplanes operating from Holloman, Cannon, and Davis-Monthan Air Force Bases and Biggs Army Airfield (Jim Eckles, "Border Star 85 Underway on WSMR, Fort Bliss," *Missile Ranger Magazine*, Vol. 38, No. 10, March 8, 1985, p. 1).

> GALLANT EAGLE 86, "a large-scale field training exercise sponsored by the U.S. Central Command, was conducted at several locations in the western United States from 25 July to 3 August. Air support for the exercise was staged from several airfields spread throughout the southwestern United States. Ground operations were conducted by exercise forces at Fort Irwin, California, and the Marine Corps Air Ground Combat Center at Twentynine Palms, California. Approximately 35,000 military personnel were involved in the exercise, with the Army Reserve and the National Guard providing extensive combat service support during all phases. Participating Army units were Third Army Headquarters; XVIII Airborne Corps Headquarters; elements of the 82d and 101st Airborne Divisions, and of the 24th ID; the 197th Infantry Brigade; 75th Infantry Regiment Headquarters; 5th Special Forces Group, 1st Special Forces; and elements of the 75th Ranger Regiment. GALLANT EAGLE 86 provided a simulated combat environment for training, planning, and execution of joint military operations. The exercise permitted an evaluation of Central Command Headquarters and portions of its multi-service forces in tactical operations in a desert setting" (Gough, 1995, p. 37).

> GALLANT EAGLE 88 involved about 30,000 soldiers, sailors, airmen, and Marines. "Supply operations constituted a major part of the training. . . . Elements of the 377th Theater Army Command and support units tested supply operations by transferring food, fuel, and spare parts from the beaches at Camp Pendleton, California, to units fighting 200 miles away in the Californian desert at 29 Palms" (William Joe Webb, *Department of the Army Historical Summary: Fiscal Year 1988*, Washington, D.C.: Center of Military History, U.S. Army, 1993, p. 49).

impacts. Although this is not necessarily a significant constraint, this likely prevented the brigade headquarters from rehearsing certain missions or functions in a manner that would match the realism of an actual operation. In addition, it seems likely that not all of the Army's brigade headquarters were able to participate in these types of events, although many different brigade headquarters did participate over time. As noted above, we were not able to make a detailed assessment of the number and types of supporting brigade headquarters that participated in the various large-scale exercises over time, nor to assess the number and size of subordinate formations they had the opportunity to command. Similarly, we were not able to assess the degree to which brigade headquarters were able to focus on their training objectives, or to determine if they received EXEVALs of their training proficiency as a result of their participation. Nonetheless, these large-scale exercises at least provided participating supporting brigade headquarters some opportunity to train and conduct mission command in a complex field environment that allowed movement and maneuver of substantial subordinate formations at a pace and scale that they might never experience elsewhere and required coordination and integration as part of a division- or corps-level combined arms team.

Between the late 1980s and mid-1990s, these large-scale FTXs were either discontinued outright or downsized into smaller exercises focused on CPXs supported by computer simulation (similar to the BCTP). The smaller exercises were sometimes combined with field training events focused on lower echelon forces, which interacted with the brigade and above headquarters participating in the CPX.[10] Reasons for this transition away from large-scale exercises included the end of the Cold War in Europe, political considerations related to negotiations with North Korea, budget and force structure reductions, environmental concerns, and an apparent conviction that computer-aided CPXs could provide comparable training value to the large-scale field exercises they replaced. Maneuver brigades could still benefit from realistic tactical training at the dirt CTCs, as well as from participation in the new computer-aided CPXs. Supporting brigade headquarters, on the other hand, lacked dirt CTC–like opportuni-

---

[10]   For example: "The emphasis on curtailing large exercises postponed REFORGER 89 to 1990 and resulted in both a substantial reduction in people and equipment and heavy reliance on computer simulations for REFORGER 90. USA-REUR developed a concept, the REFORGER Enhancement Program (REP), that allowed field training, command post, and command field exercises to run concurrently by the use of simulations. With 56,000 U.S. personnel, 15,000 of them from CONUS, REFORGER 90 took place from late December 1989 to late February 1990. The maneuver phase, named CENTURION SHIELD, pitted the V Corps against the VII Corps, and focused on training staffs, from battalion through corps, primarily by simulation. The Distributed War Game System (DWS) orchestrated operations, while the Joint Exercise Simulation System (JESS) did tactics. Two brigades of the 10th Mountain Division performed most of the actual field maneuvers of CENTURION SHIELD. In two other new developments used in REFORGER 90, U.S. forces employed no tanks and combined the use of umpires with computer calculations for adjudicating the outcome of battles" (William Joe Webb, Charles Anderson, Dave Hogan, Dale Andrade, Thomas Popa, Mary Gillett, Rebecca Raines, Glen Hawkins, and James Yarrison, *Department of the Army Historical Summary: Fiscal Years 1990 and 1991*, Washington, D.C.: Center of Military History, U.S. Army, 1997, p. 68). The last REFORGER was held in 1993. The last BORDER STAR we identified was in 1985 and the last GALLANT EAGLE we identified was in 1988.

Regarding TEAM SPIRIT:

> From 1991 to 1996, Team Spirit became both a carrot and a stick during US negotiations with North Korea over its burgeoning nuclear program. This exercise was cancelled in 1992, carried out again in 1993, and planned but not executed from 1994 to 1996 as a result of negotiations that led to the 1994 U.S.-DPRK Agreed Framework and efforts to ensure the framework remained in effect. Team Spirit was then replaced with a command post exercise known as "Reception, Staging, Onward Movement, and Integration" (RSO&I)—which was conducted from 1994 until 2007 along with "Foal Eagle," a series of tactical level exercises taking place in the spring. These exercises were much smaller than Team Spirit but maintained staff readiness in the conduct of flowing US forces to the peninsula (Robert Collins, "A Brief History of the US-ROK Combined Military Exercises," *38north.org*, February 26, 2014).

ties. By the latter 1990s, field training for these brigade headquarters was limited to what could be accomplished at home station, or as part of now much smaller-scale joint and combined exercises.

Training Opportunities Prior to the Start of ODS and OIF

To state the obvious, a key purpose of the CTC program and the Army's participation in joint and combined exercises was to prepare Army forces for LSCO. Although the peacetime training opportunities afforded supporting brigade headquarters might have been secondary to those afforded to maneuver brigades and division and corps headquarters, many of the supporting brigade headquarters that participated in ODS (1991) and OIF (2003) received an additional opportunity. In each case, as part of the immediate preparation for combat operations, at least some deploying brigade headquarters were able to participate in live, large-scale training events and exercises—in some cases, before deployment and in some cases after—that helped prepare them to operate and exercise mission command in LSCO.

In ODS, for example, XVIII Airborne Corps was the first Army corps to deploy to Saudi Arabia, arriving months before the start of the air campaign that commenced in January 1991. Its elements had substantial time to train and prepare in theater. Although specific details are unclear as to how individual divisions and brigades trained, at a minimum, the scale and complexity of training opportunities must surely have surpassed those available to supporting brigade headquarters during peacetime. As one Army historian summarized overall XVIII Corps training opportunities:

> The XVIII Corps' training priorities shifted to maneuver warfare and force modernization during December and early January. As modernization picked up speed, the corps devoted considerable time to live fire exercises to become familiar with new equipment. Shortages of training ammunition complicated such exercises, but by the start of the offensive practically every unit, except the air defense battalions, had tested its weapon systems. Often using live ammunition, corps troops worked on maneuver techniques, stressing the use of helicopters, concealment, and responses to different situations. Through exercises with the artillery and the Air Force, they improved coordination of fire support. At higher levels the corps used the Battle Command Training Program, designed at Fort Leavenworth, to hold seminars and exercises for instruction of staffs in command and control, a major concern of Lt. Gen. Gary E. Luck, the corps commander.[11]

VII Corps was the second Army corps to deploy for ODS. Receiving notification in November 1990, VII Corps and its supporting elements arrived in Saudi Arabia in piecemeal fashion starting in December, with some units arriving more than a week after the air campaign began on January 17, 1991. Nonetheless, in part given the length of the air campaign, many units still had several weeks to conduct precombat training in theater before the start of the ground campaign. In addition, VII Corps commander Lieutenant General Fred Franks used the corps' movement from its tactical assembly areas in preparation for the start of the ground campaign as a major training event. As an Army history described it:

---

[11] Frank N. Schubert and Theresa L. Kraus, eds., *The Whirlwind War: The United States Army in Operations DESERT SHIELD and DESERT STORM*, Washington, D.C.: Center of Military History, United States Army, 1995, p. 150.

With three days to conduct the move, Franks had enough time to practice his movement as well as to move into his attack position. . . . The corps could conduct an unprecedented full-scale practice of the movement to contact, with actual forces on ground almost identical to where they would soon fight. The commanders and staffs could actually test their command and control equipment and procedures. If they discovered problems, they had some time to make corrections prior to battle. . . . Division commanders were thankful that they had been able to practice moving their entire commands, with all their attached engineers, artillery, and other support. They now understood firsthand how much fuel they would consume and how long it took to refuel. They also gained a great appreciation for how much terrain they would occupy in the desert. General Smith remarked that he "had no idea how his division would fit in his [assigned] space" until the rehearsal.[12]

V Corps was the single Army corps committed to OIF. Prior to their deployment to the Persian Gulf, elements of V Corps deployed to Poland and conducted VICTORY STRIKE, which was partly an FTX focused on using deep fires and attack aviation; the corps also used this as a deployment exercise.[13] This was followed by VICTORY SCRIMMAGE, a BCTP-supported CPX at which representatives of all subordinate divisions and separate brigades were present.[14] Although VICTORY SCRIMMAGE was not an FTX, it allowed all major commands under V Corps an opportunity to rehearse the plan and refine SOPs prior to the start of combat operations. In addition, V Corps Support Command

conducted a weeklong rehearsal of the entire range of logistics efforts required by the vast distances, large formations, and major combat operations of the coming campaign. The logistics rehearsal identified a number of challenges that logisticians were able to adapt to during deployment and before the beginning of hostilities.[15]

Because of the timing of the OIF force flow in early 2003, the 3rd ID was the only Army division that was fully deployed well in advance of the start of LSCO (the 101st Airborne Division arrived in theater approximately ten days before the start of combat operations, and the 4th ID's equipment remained afloat, bound for Kuwait after the Turkish government refused to allow it to stage from Turkey). Along with the 101st Airborne Division, the 3rd ID also experienced the most sustained combat before the end of LSCO. Once in theater, the 3rd ID had the opportunity to conduct substantial precombat training. As an Army history described it:

Although it was far away from their families and the comforts of home, the Kuwaiti desert offered vast training space. Moreover, with the Army gearing up for combat, the usually scarce training resources—ammunition, time, and fuel—were abundant. . . . Brigades began training in earnest, using the Udairi Range complex and the vast expanse of desert to practice offensive operations. Training right up until they attacked into Iraq, during the next four months the division would fire, drive, and fly the equivalent of two years of train-

12   Stephen A. Bourque, *Jayhawk! The VII Corps in the Persian Gulf War*, Washington, D.C.: Center of Military History, U.S. Army, 2002, pp. 172, 178.

13   Gregory Fontenot, E. J. Degen, and David Tohn, *On Point: The United States Army in Operation IRAQI FREEDOM*, Fort Leavenworth, Kan.: Combat Studies Institute Press, 2004, p. 53.

14   Fontenot, Degen, and Tohn, 2004, p. 55.

15   Fontenot, Degen, and Tohn, 2004, p. 55.

ing ammunition and fuel, roughly six times what they would have experienced in peacetime. This precious training opportunity, afforded only to 3rd ID because of the unique buildup in this campaign, contributed significantly to the division's success.[16]

In both ODS and OIF, at least some deploying brigade headquarters were able to participate in live, large-scale training events and exercises that helped prepare them to operate and exercise mission command in LSCO. In the future, there might be cases in which the Army can again benefit from similar opportunities to engage in substantial precombat training. However, there are likely to be other cases where such opportunities will not exist—for example, in rapidly unfolding contingencies that occur with little or no warning. In such cases, the quality of peacetime training will go a long way toward setting the conditions for success or failure in first battles.

---

[16] Fontenot, Degen, and Tohn, 2004, pp. 58, 76.

Sample Interview Questions

Interviews with key stakeholders contributed significantly to the findings and recommendations in this report. Our interviews focused on the following groups of stakeholders:

- MCTP
- various CoEs (including the Mission Command CoE) and Branch Schools that develop doctrine and training products
- leaders and/or key staff members from certain F/MF brigade headquarters.

This appendix provides samples of the interview questions we prepared in advance of our interviews. Not all questions were asked in all cases, and the lists do not include any follow-up questions that arose in the context of specific interviews.

MCTP

1. What role does MCTP play in training units to conduct mission command? How does this vary by unit type?
2. What types of training does MCTP provide to units?
   a. How was this training developed? Based on what guidance?
   b. How does this vary by unit type?
2. Do you provide any mission command training guidance to CoEs or branch schools?
3. In your opinion, is this training sufficient to prepare units to conduct mission command? How does this vary by unit type?
4. What are the strongest/weakest parts of current mission command training? How does this vary by unit type?
5. What are the most significant challenges for mission command training? How does this vary by unit type?
6. What gaps exist in mission command training? How could these gaps be addressed? What challenges do these gaps cause?
7. Do specific types of units perform better than others in MCTP activities or exercises? If so, what kinds of units? Do you have any sense as to why this might be?
8. Do you collect any data on mission command training performance? What sorts of analysis do you do with this data?

CoEs and Branch Schools

1. In an LSCO environment, how does [brigade type X] execute mission command given its specific mission and capabilities?
   a. Which key higher and parallel organizations does it need to synchronize with?
   b. What are typical command and support relationships with subordinate units?
   c. Can you comment on risks associated with having to operate on short notice in an LSCO environment with units that it has not habitually trained and integrated with?
2. What are key doctrine and training products?
   a. Do you think there are any gaps or areas for improvement?
   b. Mission command is a supporting collective task to each brigade MET. In developing training products, how do you attempt to integrate mission command and the operations process into MET-focused products?
3. How do soldiers receive greatest exposure to doctrine on mission command and the operations process (e.g., professional development courses, learn on the job, etc.)?
4. What are the greatest challenges for [brigade type X] in training to execute mission command and the operations process? Consider:
   a. lack of habitual relationships with higher, parallel, or subordinate echelons
   b. lack of opportunities to train with UAPs.
5. In your opinion, are current training approaches sufficient to prepare [brigade type X] to conduct mission command?
6. Do you think there has been a loss of institutional memory on what performing mission to standard in LSCO looks like?
   a. If so, does this impact the Army's ability to grade itself as to whether it is on the right track in getting back to mission command proficiency for LSCO?
7. Do you collect any data on mission command training lessons/best practices?
   a. Do you receive feedback from any of the following sources: EXEVALs, WFXs, other exercises (e.g., joint exercise program), home station training?
   b. Is the feedback mechanism formal/systematic or informal/anecdotal?

F/MF Brigade Headquarters

1. Overall, in the context of performing mission command in an operational environment, how do the following actions present challenges for your unit:
   a. integration with parallel and higher echelons
   b. exercise of mission command over subordinate elements
   c. coordination with UAPs.
2. How are mission command doctrine and training products incorporated into unit training plans and evaluations?
   a. What are strengths and limitations of current doctrine and training products?
3. How well do home station training events, WFXs, and other training opportunities support mission command training?
   a. What does your standard home station training look like in an average year?
   b. If your unit recently completed (or is preparing for) a WFX:

       i. How did your unit prepare to attend the WFX? What sorts of home station training did you complete that was specific to WFX preparation?

      ii. How well does the WFX simulate a real-world environment for LSCO?

     iii. Does the fact that the WFX is focused primarily at corps and division headquarters significantly impact the effectiveness of the WFX as a training opportunity for your unit?

  c. If your unit did not (or will not) participate in a WFX, are there other major training opportunities that serve as a substitute?

  d. How could WFXs or other major training opportunities be improved to help build mission command proficiency?

4. How does the level of synchronization with and support from division/corps headquarters impact the quality, resourcing, and outcomes of mission command training, whether routine home station training or for a WFX or other major training event?

5. How well are EXEVALs used to validate mission command proficiency?

  a. Has your unit received an EXEVAL? When and by whom was it performed? How did it impact your unit's training thereafter?

6. Overall, what are the greatest challenges your unit faces in training for and/or exercising mission command, and do you have recommendations for mitigating them?

References

196th Infantry Brigade, "Joint Pacific Multinational Readiness Capability," webpage, undated. As of July 11, 2019:
https://www.usarpac.army.mil/196th/JPMRC/index.asp

ADP—*See* Army Doctrine Publication.

ADRP—*See* Army Doctrine Reference Publication.

Army Doctrine Publication 5-0, *The Operations Process*, Washington, D.C.: Headquarters, Department of the Army, May 17, 2012.

Army Doctrine Publication 6-0, *Mission Command*, Change No. 2, Washington, D.C.: Headquarters, Department of the Army, March 12, 2014.

Army Doctrine Reference Publication 2-0, *Intelligence*, Change No. 2, Washington, D.C.: Headquarters, Department of the Army, September 4, 2018.

Army Doctrine Reference Publication 3-09, *Fires*, Change No. 1, Washington, D.C.: Headquarters, Department of the Army, February 8, 2013.

Army Doctrine Reference Publication 4-0, *Sustainment*, Washington, D.C.: Headquarters, Department of the Army, July 13, 2012.

Army Doctrine Reference Publication 5-0, *The Operations Process*, Washington, D.C.: Headquarters, Department of the Army, May 17, 2012.

Army Doctrine Reference Publication 6-0, *Mission Command*, Change No. 2, Washington, D.C.: Headquarters, Department of the Army, March 28, 2014.

Army Techniques Publication 2-19.3, *Corps and Division Intelligence Techniques*, Change No. 1, Washington, D.C.: Headquarters, Department of the Army, March 10, 2016.

Army Techniques Publication 3-04.1, *Aviation Tactical Employment*, Washington, D.C.: Headquarters, Department of the Army, April 13, 2016.

Army Techniques Publication 3-09.24, *Techniques for the Fires Brigade*, Washington, D.C.: Headquarters, Department of the Army, November 21, 2012.

Army Techniques Publication 3-09.90, Division Artillery Operations and Fire Support for the Division, Washington, D.C.: Headquarters, Department of the Army, October 12, 2017.

Army Techniques Publication 3-34.23, *Engineer Operations: Echelons Above Brigade Combat Team*, Washington, D.C.: Headquarters, Department of the Army, June 10, 2015.

Army Techniques Publication 4-93, *Sustainment Brigade*, Washington, D.C.: Headquarters, Department of the Army, April 11, 2016.

Army Training Network, *METL Viewer*, undated. As of December 24, 2019:
https://atn.army.mil/ATNPortalUI/METL

ATP—*See* Army Techniques Publication.

Battle Command Training Program, "Ops Grps Sierra and Foxtrot: Training Support/Functional/Theater Bdes," briefing slides, undated.

BCTP—See Battle Command Training Program.

Bourque, Stephen A., *Jayhawk! The VII Corps in the Persian Gulf War*, Washington, D.C.: Center of Military History, U.S. Army, 2002.

Cocke, Karl E., *Department of the Army Historical Summary: Fiscal Year 1985*, Washington, D.C.: Center of Military History, U.S. Army, 1995. As of January 22, 2019:
https://history.army.mil/books/DAHSUM/1985/index.htm

Collins, Robert, "A Brief History of the US-ROK Combined Military Exercises," *38north.org*, February 26, 2014. As of January 22, 2019:
https://www.38north.org/2014/02/rcollins022714

Eckles, Jim, "Border Star 85 Underway on WSMR, Fort Bliss," *Missile Ranger Magazine*, Vol. 38, No. 10, March 8, 1985.

Field Manual 3-0, *Operations*, Washington, D.C.: Headquarters, Department of the Army, October 6, 2017.

Field Manual 3-04, *Army Aviation*, Washington, D.C.: Headquarters, Department of the Army, July 29, 2015.

Field Manual 3-09, *Field Artillery Operations and Fire Support*, Washington, D.C.: Headquarters, Department of the Army, April 4, 2014.

Field Manual 3-34, *Engineer Operations*, Washington, D.C.: Headquarters, Department of the Army, April 2, 2014.

Field Manual 6-0, *Commander and Staff Organization and Operations*, Washington, D.C.: Headquarters, Department of the Army, May 11, 2015.

Field Manual 6-02, *Signal Support to Operations*, Washington, D.C.: Headquarters, Department of the Army, January 22, 2014.

Field Manual 7-0, *Train to Win in a Complex World*, Washington, D.C.: Headquarters, Department of the Army, October 5, 2016.

FM—*See* Field Manual.

Fontenot, Gregory, E. J. Degen, and David Tohn, *On Point: The United States Army in Operation IRAQI FREEDOM*, Fort Leavenworth, Kan.: Combat Studies Institute Press, 2004.

Gallahue, Kimo, "Mission Command Training Program Overview Brief," briefing, February 10, 2017.

Gough, Terrence J., *Department of the Army Historical Summary: Fiscal Year 1986*, Washington, D.C.: Center of Military History, U.S. Army, 1995. As of January 22, 2019:
https://history.army.mil/books/DAHSUM/1986/index.htm

Schubert, Frank N., and Theresa L. Kraus, eds., *The Whirlwind War: The United States Army in Operations DESERT SHIELD and DESERT STORM*, Washington, D.C.: Center of Military History, U.S. Army, 1995.

TRADOC—See U.S. Army Training and Doctrine Command.

U.S. Army Combined Arms Center, "FM 7-0 Tutorial: Overview and How to Use the Training & Evaluation Outline (T&EO)," briefing, March 22, 2017.

U.S. Army Directorate of Force Management, *FMSWeb*, database, undated. As of December 24, 2019:
https://fmsweb.fms.army.mil

U.S. Army Training and Doctrine Command, *Mission Command Training Program*, Fort Eustis, Va., TRADOC Regulation 350-50-3, June 23, 2014.

U.S. Army Training and Doctrine Command, *The U.S. Army in Multi-Domain Operations 2028*, Fort Eustis, Va., TRADOC Pamphlet 525-3-1, November 27, 2018. As of November 4, 2019:
https://www.tradoc.army.mil/Portals/14/Documents/MDO/TP525-3-1_30Nov2018.pdf

Wascom, Charles L., "5th Infantry Division," *Infantry: A Professional Journal for the Combined Arms Team*, Vol. 69, No. 3, May–June 1979.

Webb, William Joe, *Department of the Army Historical Summary: Fiscal Year 1988*, Washington, D.C.: Center of Military History, U.S. Army, 1993. As of January 22, 2019:
https://history.army.mil/books/DAHSUM/1988/index.htm

Webb, William Joe, Charles Anderson, Dave Hogan, Dale Andrade, Thomas Popa, Mary Gillett, Rebecca Raines, Glen Hawkins, and James Yarrison, *Department of the Army Historical Summary: Fiscal Years 1990 and 1991*, Washington, D.C.: Center of Military History, U.S. Army, 1997. As of January 22, 2019:
https://history.army.mil/books/DAHSUM/1990-91/index.htm